Molecular Mechanisms of Autonomy in Biological Systems

Tara Karimi

Molecular Mechanisms of Autonomy in Biological Systems

Relativity of Code, Energy and Mass

Springer

Tara Karimi
Tulane Medical Center
Tulane University
New Orleans, LA, USA

ISBN 978-3-030-06304-7 ISBN 978-3-319-91824-2 (eBook)
https://doi.org/10.1007/978-3-319-91824-2

Printed on acid-free paper

This Springer imprint is published by the registered company Springer International Publishing AG part of Springer Nature.
The registered company address is: Gewerbestrasse 11, 6330 Cham, Switzerland

To my parents and teachers with love
To my brother Moji, for his support and passion for science

Preface

From my background in biomedical science, I have always been fascinated by endogenous intelligence and the autonomous capabilities of biological systems.

I remember from my courses in basic sciences including chemistry and physics as well as biochemistry and biophysics that the definition of life was so mysterious. I have several questions about the basic principles of life; for instance, what is the logic and law behind life? Are all natures' laws in chemistry and physics applicable in living systems as well? Which properties of living systems distinguish them from other nonliving elements of nature?

The curiosity about the life motivated me to continue my studies in different areas of life sciences including biochemistry and later stem-cell-based regenerative medicine, genetics, and molecular biology.

After I started my Ph.D. program in biochemistry, I realized that if we delve deep enough in a living cell, we see that a cell is merely a chemical system composed of all normal elements of nature (remarkably enriched with five elements including carbon, hydrogen, oxygen, nitrogen, and phosphorous). However, the chemistry of a living cell has been designed in a special way by nature to function as an autonomous system. In fact, the chemistry of a cell functions as an algorithm at the same time, and this makes the chemistry of a living cell special.

Different forms of living systems, from the most primitive cells to the complex multicellular organisms, illustrate special properties of highly advanced autonomous systems including self-regulation, self-assembly, self-organization, self-healing, self-replication, and endogenous intelligence, cognition, and information heritage among generations. Still, living cells are in close contact with nonliving elements of nature. In fact, living systems are highly integrated with nonliving elements of nature.

From that point, the entire natural system became a large puzzle in my mind. I could see the similarities and overlaying areas in the borders of different pieces of this puzzle, but I needed to arrange them beside each other appropriately. While arranging the pieces of this puzzle of nature's law (a mix of biological systems and the nonliving elements), I decided to investigate the matching pieces by zooming into the borderlines. Therefore, I started to search among different branches of basic sciences (including chemistry, physics, and mathematics).

During this journey, I realized that nature does not separate chemistry, physics, and mathematics from each other; everything happens in harmony. We, humans, made these classifications to make it easier to be studied. It is like all geographic borders and lines that we humans agreed to draw between countries and cities on a map. However, if we look at the satellite view of Earth from space, we realize that there is no line between different parts of Earth. Nature does not isolate life from chemistry and physics, but all different parts of nature are highly integrated to each other while all are regulated by the same law.

My curiosity for an understanding of the natural laws regulating biological systems pushed me harder to study different areas of science, looking for the missing pieces in my puzzle. Finally, I found that the missing part is the coding system that can be integrated into the electrochemical interactions between molecules and atoms. Molecules with this natural coding system for integration of information, can handle an operation more than simple storage of energy in their chemical bonds of molecules.

Our current understanding of nature's laws in physics and chemistry is limited to the relativity of energy and mass while missing code as the third dimension of nature. By considering the possibility of information storage in molecules as an additional aspect of nature's chemical and physical laws, we can provide a clear description for all autonomous properties of biological systems including self-regulation, self-organization, self-regeneration, endogenous intelligence, and cognition. In addition, by applying the basic principles of the relativity of code, energy, and mass, it is possible to design synthetic systems with similar properties of biological systems for different industrial applications. Here, by learning from nature and considering the possibility of information storage and operation by coded molecules, as well as programmable interaction among molecules with each other and their environmental factors, I discuss a new emerging concept as *cognitive chemistry*.

Cognitive chemistry is a special form of chemistry where the system is able to recognize and respond to the dynamic environmental signals and even get adapted to its environmental conditions. The system is highly connected to its environmental factors through molecular sensors and logic gates. The system is made of a cluster of precisely coded molecules (that through their electrochemical interactions) are highly connected to each other as well as their environmental factors. In fact, a cognitive chemistry system has the capacity of endogenous cognition.

In this book, I attempted to connect the dots between different branches of natural sciences. To this end, I tried to explain the molecular mechanism of autonomy in biological systems in a simple language to be applicable for researchers in different fields of sciences (other than biology) including computer science, physics, chemistry, and material science. I highly recommend the audience to see references at the end of each chapter for the detailed information.

In each chapter of this book, I discussed a specific aspect of autonomy in biological systems (e.g., self-regulation, self-fueling, and self-organization) through the relativity of code, energy, and mass. In addition, I discussed how relativity of code, energy, and mass can be applied in different industries.

In a separate chapter in this book, I discussed how learning from the molecular regulation of autonomy and self-organization in biological systems can be applied for innovative therapeutic approaches in the treatment of diseases with no current cure applying the conventional methods in biomedical sciences.

I attempted to connect the dots between different dimensions of natural sciences to make a new image of nature's laws which involves additional dimensions. To that end, I highly recommend studying the different chapters of this book in order.

New Orleans, LA, USA
January 2018

Tara Karimi

Contents

Chapter 1
The Relativity of Code, Energy, and Mass Versus the Relativity of Energy and Mass

Abstract As a puzzling feature of nature, biological systems possess several aspects of autonomy including self-organization, self-replication, self-fueling, and self-regulation. It is initially difficult to accept the idea that each living creature is merely a chemical system. Yet, scientifically, biological systems are mainly composed of four natural chemical elements, carbon, hydrogen, oxygen, and nitrogen. Considering that biological systems are highly connected to the other inanimate elements in nature, several challenging questions remain to be answered;

- What are the regulatory mechanisms behind the autonomous properties of biological systems that distinguish them from inanimate elements and non-living systems in nature?
- Can we explain all autonomous properties of biological systems applying the currently defined natural laws of physics and chemistry?

Current understanding of natural sciences is limited to the relativity of energy and mass while missing code (embedded information in molecules) as the third dimension of nature's law. Here, in this chapter, we first discuss the foundation of chemistry and physics in both non-living and living systems. Then, we introduce code as the third dimension of nature's law, through the capacity of information storage in molecules (coding capacity of biomolecules). Considering the coding capacity of biomolecules, as an additional factor beyond the energy storage in molecules, we define a new concept of the relativity of code, energy, and mass. The relativity of code, energy, and mass as a new platform in science, unveils the molecular logic behind several unknown features of nature, such as autonomous properties of biological systems.

Keywords Algorithmic Logic of life · Origin of life · Chemistry of Life · Autonomy in Biological Systems · Self- assembly · Self-organization · Self-fueling · Self-replication and Self-regeneration. Relativity of Energy and Mass · Relativity of Code · Energy and Mass

T. Karimi, *Molecular Mechanisms of Autonomy in Biological Systems*,
https://doi.org/10.1007/978-3-319-91824-2_1

1.1 Introduction

The current understanding of basic natural sciences (including physics and chemistry) is focused on energy and mass as the main aspects of any matter (material) in the universe. The correlation between energy and mass as two major aspects of matter has been described clearly through the equation of relativity of energy and mass by Albert Einstein. However, still, the possibility of information storage in molecules (embedding code in molecules) remains to be described systematically as the third dimension of matter. Biomolecules such as nucleic acids are promising models for the description of the possibility of information storage in the matter (coding capacity of matter).

Here, in this chapter, we first discuss the foundation of chemistry and physics in both nonliving and living systems. Then, by referring to the possibility of information (code) storage in molecules, we introduce code as the third dimension of matter (e.g., embedded codes in biomolecules).

By considering the information storage capacity of biomolecules, we describe this three-dimensional functional property of biomolecules (and any other programmable molecules) by the relativity of code, energy, and mass.

In fact, by learning from autonomous properties of biological systems and recapitulating of the coding capacity of biomolecules, we define a new translation of relativity equation as the relativity of code, energy, and mass.

The relativity of code, energy, and mass as an emerging concept in natural sciences unveils the molecular logic behind several features of nature, including the autonomous properties of biological systems.

Beyond the biological sciences, application of basic principles of the relativity of code, energy, and mass can open new avenues toward the generation of synthetic autonomous systems with a wide range of applications in different areas of science including computer, energy, physics, and material sciences.

1.2 Biological Systems Are Highly Autonomous

Living organisms are highly advanced autonomous systems. By definition, a key feature of autonomy in both natural and artificial systems is related to the ability of independent operation of processes [11, 15].

Based on the theory of cosmic cataclysm, which is the most popular theory for the origin of the universe, about 10–20 billion years ago, the universe formed as a massive cataclysmic eruption of hot and energy-rich subatomic particles. The theory maintains that after the eruption, the universe expanded. As time passed and matter cooled, more diverse types of atoms began to form and they eventually condensed and produced stars and galaxies of our present universe, including earth and all its chemical elements [12, 20]. Even though the origin of life cannot be precisely detected, based on the evidence and residuals of bacteria-like organisms found from geological research, it is estimated that life initiated on earth about 3.5 billion years ago. Even earlier, almost 4 billion years ago, very simple components of life such as primitive functional molecules existed. Over time, these functional molecules connected to each other and formed primitive forms of life [9]. Biology is the natural science that involves the study of physical, chemical, functional, and developmental properties of life. Living organisms (also called biological systems) are considered as highly complex autonomous systems [9, 17].

As a remarkable feature of nature, biological systems possess several aspects of autonomy [15]:

1. Biological systems are self-assembling.

Self-assembly is a process in which a disordered system with many components turns into an orderly and stable structure (at equilibrium with minimum energy configuration) without external direction. It usually involves only the weaker bonds (e.g., hydrogen bonds and hydrophobic interactions), which makes the structure of the system more flexible.

An example of self-assembly in non-living systems is the formation of crystals. The folding of polypeptide chains into proteins and nucleic acids into their functional forms are examples of self-assembled biological structures.

2. Biological systems are self-organizing.

Self-organization means a system has the capacity to assemble its own subunits through multiple layers and scales of complexity. Embryonic development is a typical example of self-organizing processes in biological systems [13]. During the early stages of embryonic development, stem cells are able to assemble and make highly specific structures which grow in size and complexity over time [8, 19] (Figs. 1.1 and 1.2).

Even though both the self-assembly and self-organization processes refer to the spontaneous formation of orderly structure, they can be distinguished from each other in different aspects. While self-assembly arrives at an equilibrium configuration, the self-organization drives the system away from equilibrium which requires an infusion of energy to maintain. Another major difference is the interaction. While the interaction in self-assembly is more elementary, which can be traced back to the electromagnetic force, the interaction in self-organization is more complex. The size of the resulting structure from self-assembly is mostly mesoscopic, but it can be considerably larger for self-organization [13].

To summarize, self-assembly can be observed in both living and nonliving systems, while self-organization is a special property of living systems.

3. Biological systems are self-replicating.

Self-replication means a system can produce a precise copy of itself. All living systems, regardless of their developmental level – from the most primitive bacteria to the more developed eukaryotic organisms – can make a faithful copy of themselves.

Fig. 1.1 (**a**, **b**) Images illustrating the embryonic self-organization in a chicken embryo which leads to the autonomous formation of complex structures from less organized materials by time

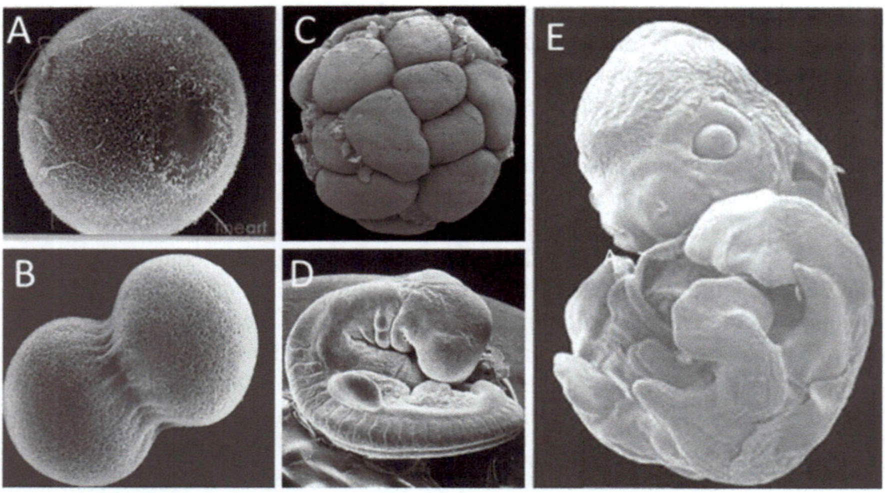

Fig. 1.2 (**A**-**E**). Schematic image representative of self-organization of stem cells during the early stages of embryonic development. Stem cells are able to assemble and make highly specific structures, which grow in size and complexity by time. (**A**-**E**) Scanning micrographs of the human embryo at different stages of development (Carnegie stages). (Images are reproduced with the permission of Dr. Kathy Sulik, from the collection of Dr. Vekemans and Tania Attie-Bitach. Source of images: https://embryology.med.unsw.edu.au/embryology/index.php/Carnegie_stage_table)

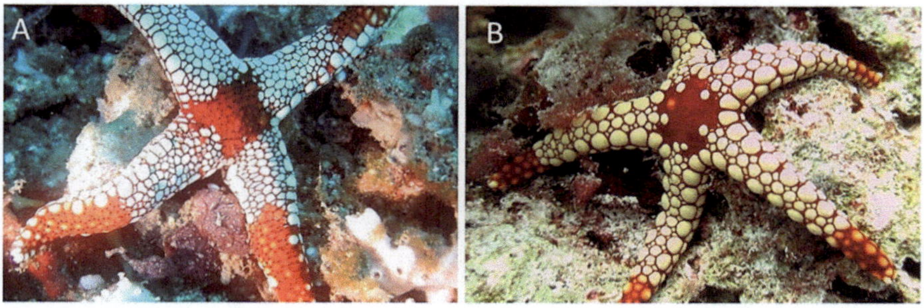

Fig. 1.3 Self-healing capacity of starfish (*Fromia Monilis*) to repair the damaged arm. (Source of image https://www.whatsthatfish.com/fish/necklace-sea-star/327#2)

4. Biological systems are self-healing.

Self-healing means a system has the capability to directly regenerate its own damaged part by itself, independent of any external mediator or operator. For example, Fig. 1.3 illustrates the self-healing capacity of starfish and complete regeneration of a damaged arm.

5. Biological systems are self-fueling.

Self-fueling means a system has the capability to directly produce and uptake its own energy requirements from its environment, without the need for any external mediator or operator [12].

6. Biological systems indicate machinery behavior at multiple layers of complexity.

In addition to the self-assembly and self-organization which lead to the formation of highly specific structures, thousands of machineries can be formed during the progress of a self-organizing process in biological systems.

Unlike any other human-made systems, biological systems are made of millions of machineries at different scales, ranging from nano- to micro- and meso-machineries, which are embedded together to form functional modules of a biological system. Figure 1.4 (a–g) is a schematic image representative of biological machineries ranging from molecular level (e.g., DNA polymerase) to cellular organelles (e.g., ribosome as a part of protein synthesis machinery), tissue and organs (e.g., highly advanced renal filtration machinery), etc.

7. Biological systems are self-controlling in their interaction with environmental factors.

Self-controlling means a system has the capability to code, control, and run its own processor [12]. Biological systems apply very specialized molecular sensors and receptors for the dynamic interaction with their environmental factors. Figure 1.5 illustrates an example of self-controlling behavior in a biological system includ-

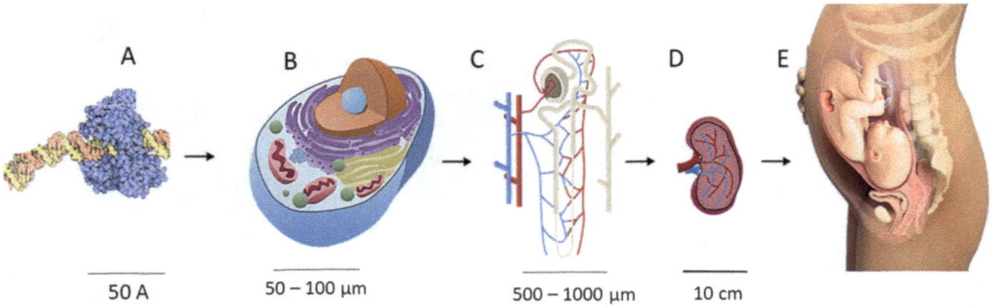

Fig. 1.4 (**a–g**) A schematic representation of the existence of multiple layers of machinery in biological systems. The images (**a–g**) provide examples of biological machinery at a wide range of scales: from molecular to micro- and macro-levels. (**a**) DNA polymerase as a molecular machinery (size, 50 Å), (**b**) ribosome (size, 50 nm), (**c**) cell (10–50 µm, a nephron), (**d**) a nephron (size, 500–1000 µm), (**e**) a renal lobule (size, 0.5–1 cm), (**f**) a kidney (size, 10 cm), and (**g**) the whole human body machinery

Fig. 1.5 Self-navigation capacity of migratory birds. The existence of magnetoreceptor proteins in eyes may regulate their capacity to detect the magnetic field of the earth and use it to navigate directions over short and long destinations. (**a**) Robin. (**b**) *Alcedo Atthis*. (Source of pictures: https://pixabay.com)

ing the self-navigation capacity of birds during their migration flight. The existence of magnetoreceptor proteins in their eyes may be responsible for the capacity to detect the earth's magnetic field and use it to navigate directions over short and long destinations.

8. Biological systems have the capacity of sensing and responding to environmental factors.

The capacities of sensing and responding to the environmental factors (or signals) lead to cognitive behavior in a system and in the advanced form allow for acqui-

Fig. 1.6 (**a–c**) Biological systems are able to receive and respond to their environmental signals applying a cluster of molecular sensors and operators which lead to their cognitive behavior. (**a**) Cognitive behavior of chameleon to synchronize its skin color with the environment. (**b**) Receiving environmental signals applying molecular sensors of eyes. (**c**) Visible changes in skin pigmentation for the synchronization with the environment

sition, perception, understanding, and remembering of information [12]. Figure 1.6 illustrates the cognitive behavior of chameleon for synchronization of its skin color with its environment.

Given that biological systems are highly connected to the inanimate elements and non-living systems in nature, several challenging questions remain to be answered.

- What are the regulatory mechanisms behind the autonomous properties of biological systems that distinguish them from inanimate elements and non-living systems in nature?
- Are we able to explain all autonomous properties of biological systems by applying the currently defined natural laws in physics and chemistry?

In the following parts of this chapter, we will determine the significance and fundamental applications of regulatory mechanisms of autonomy in biological systems. In parts 3 and 4, we will explain the chemical and physical foundations of life. In part 5, we will introduce a new scientific concept which considers code as the third dimension of nature's laws and explains different features of autonomy in biological systems.

1.3 Significance and Fundamental Applications of Regulatory Mechanisms of Autonomy in Biological Systems

Deep understanding of regulatory mechanisms behind the autonomous properties of biological systems is important in several aspects including:

1. Considering human body (including all internal organs and brain) as a highly advanced autonomous system, every disease is a result of a disorder in the regulatory mechanisms and programs that are running this autonomous system.
2. Deep understanding of molecular programs behind the human body machinery is critical for the invention of novel therapeutic methods, especially approaches for currently incurable disease (e.g., cancer, genetic disorders, Alzheimer's, Parkinson's, and multiple sclerosis).
3. The increasing demand for creation of artificial intelligence (AI) systems capable of direct connection to the human brain necessitates a deep understanding of molecular mechanisms of autonomous data processing in the human neural system. This is critical for the generation of innovative intelligent algorithms that could be used for a wide range of applications. For example, simulation of mechanisms of autonomous data processing in the human brain can be applied to generate wearable AI systems (e.g., human brain-machine interfaces) for different therapeutic purposes or improvement of human brain performance.
4. Since biological systems are highly advanced natural autonomous systems, understanding and reverse engineering of their operating system can be applied for designing of synthetic intelligent systems in a wide range of industries such as smart electronic devices, smart materials, and biocompatible and wearable intelligent devices.

1.4 Chemical Foundation of Life

Initially, it might be difficult to accept the idea that each living creature is merely a chemical system [2]. Until the nineteenth century, animals were believed to contain a vital force – "an animus" – that was responsible for their distinctive properties [12]. However, scientifically, biological systems are just composed of natural chemical elements. Cells, as the main subunit of biological systems, are made of the hierarchal organization of molecules. Figure 1.7 illustrates the structural hierarchy in the cellular organization.

The chemistry of biological systems (from the most primitive forms to the highly advanced organisms) is mainly composed of four natural elements: carbon, hydrogen, oxygen, and nitrogen, normally making up to 96% of the dry mass of living cells (Fig. 1.8). Aside from the aforementioned four elements, the next key elements of biological systems are phosphorus, sulfur, sodium, chloride, potassium, calcium, and magnesium, which make up to 3.5% of cell's dry mass [12, 14, 16].

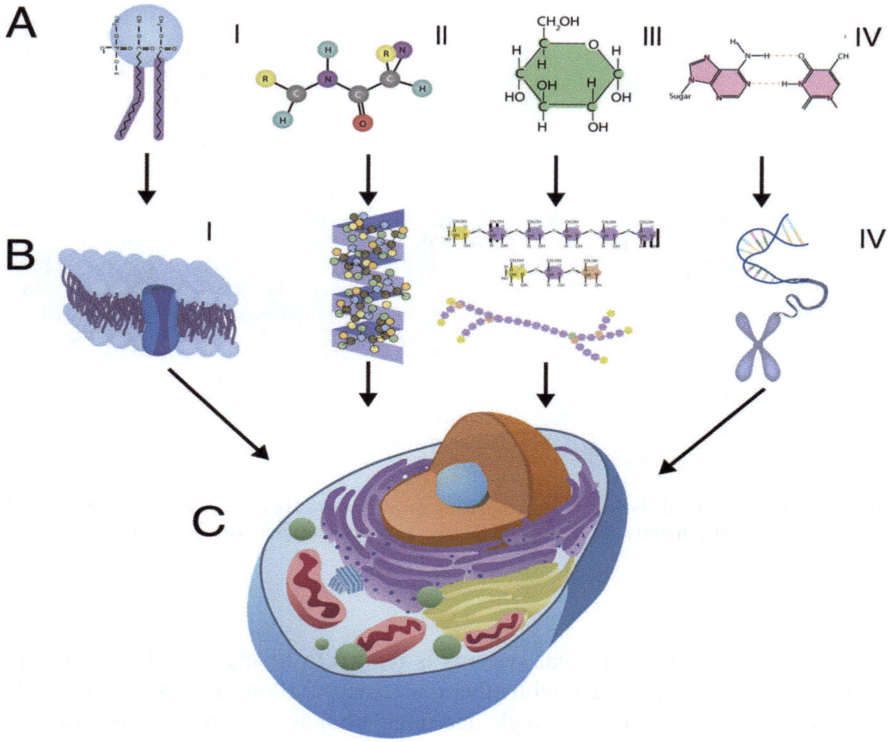

Fig. 1.7 (**a**–**c**) Schematic image representative of the chemical composition of living cells. The bottom-up assembling of molecular subunits and the formation of macromolecules, organelles, and a complete cell through multiple layers of self-organization in living cells. (**a**) Schematic representative of the basic chemical structure of nucleotides, monosaccharides, amino acids, and lipids, as the four main categories of biomolecules in living cells. (**b**) Formation of secondary and tertiary structures of macromolecules (DNA, polysaccharides, proteins, and lipid bilayer) by assembling of their molecular subunits. (**c**) Self-assembling of macromolecules and formation of cells as the basic subunits of a biological system

In addition, trace elements are present at very low levels (less than 0.5% of dry mass) in an organism. These trace elements are essential to life and living systems cannot survive without them. Trace elements include iron, iodine, manganese, molybdenum, selenium, silicon, tin, vanadium, boron, chromium, cobalt, zinc, copper, and fluorine [2]. Trace elements are essential for sustaining living organisms because of their critical role in the function of specific proteins. For example, iodine is important for making different forms of thyroid hormone and involves the regulation of growth and energy levels in mammalians. Many of the trace elements are required for the appropriate function of enzymes. Iron, which is found in red blood cells, helps transport oxygen to different tissues.

Among the four main elements of living cells, carbon makes up more than half of dry weight of cells and plays the most significant role in living organisms. The

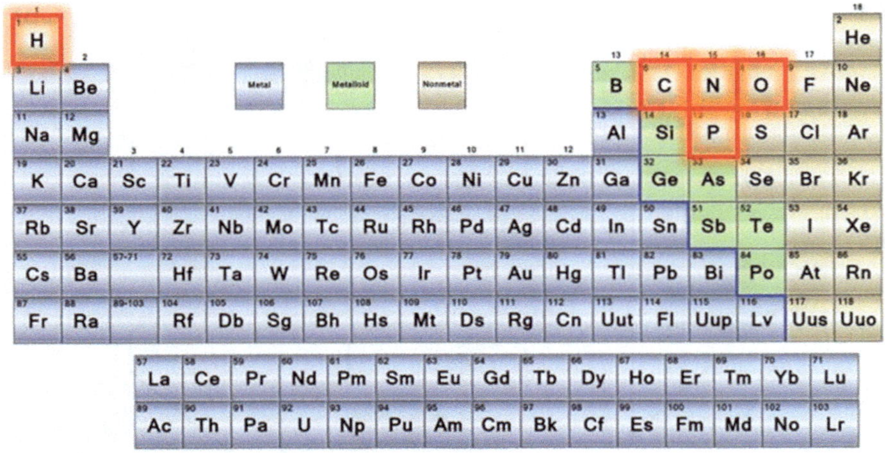

Fig. 1.8 Table of natural chemical elements. Biological systems are chemical systems, composed of natural elements, mainly including hydrogen, carbon, nitrogen, oxygen, phosphorus, and sulfur

key role of carbon in biological systems is related to its high flexibility to form chemical bonds with itself and with other molecules. Carbon can form single bonds with hydrogen atoms and both single and double bonds with oxygen and nitrogen atoms.

One of the most significant properties in biology is the ability of carbon atoms to form very stable carbon-carbon single bonds. Each carbon atom can form single bonds with up to four other carbon atoms [2]. Two carbon atoms also can form double or triple bonds. Thus, covalently linked carbon atoms can make a wide range of molecular structures including linear chains, branched chains, and cyclic structures. Groups of reactive atoms such as oxygen, nitrogen, and phosphorus, also called functional groups, are added to the main carbon skeleton to form different types of biomolecules. The existence of functional groups in the chemical structure of biomolecules determines the reactivity, solubility, and other physical properties of a biomolecule and defines their biological role in a living organism [14].

Alcohols are biomolecules that contain at least one hydroxyl group. With a strong electronegativity of oxygen atoms, hydroxyl groups provide polarity for biomolecules. Similarly, carbonyl groups in aldehydes and ketones usually increase polarity and reactivity of organic molecules. Carboxyl groups, which contain two oxygen atoms in their structure, are highly polar and reactive functional groups. Amines contain amino groups. Like free ammonia (NH_3), amines are weakly basic and can bond to electropositive hydrogen ions in a solution. Therefore, amines are able to make hydrogen bonds with water and other polar molecules in an aqueous solution. On the opposite end of the spectrum, phosphate groups are highly acidic and are mostly in an ionized state in aqueous solutions by releasing two electropositive hydrogen ions. Due to the four electronegative oxygen atoms, phosphate groups are extremely reactive. Transfer of phosphate groups among molecules can deliver energy to chemical reactions. For example, adenosine triphosphate (ATP), which contains three phosphate groups in its structure, functions as the main energy carrier in biological systems [16].

Many biomolecules are polyfunctional, containing two or more different functional groups in their chemical structures. This structural complexity of polyfunctional molecules provides a special flexibility for these molecules to shift between different roles based on the system's demand.

1.5 Physical Foundation of Life: Thermodynamics and General Relativity of Energy and Mass

Living cells must perform work to sustain the organization of their molecules, reproduce, and, in general, continue living. The synthetic reactions that occur within the cell, like the synthetic processes in any factory, require the input of energy. Energy is also consumed for motion and machinery operations of living cells. Processing and storing information also require energy, without which structures rich in information become disordered.

For chemical reactions occurring in a solution, we can define a system as all the constituent reactions and products: the solvent that contains them and the immediate atmosphere within a defined region of space. Therefore, a biological system is considered as an open system – it exchanges both energy and matter (mass) with its surroundings [12].

Biological systems derive energy from their surroundings in two ways:

1. They uptake chemical energy (such as glucose) from their environment and extract energy by their oxidization.
2. They absorb energy from sunlight.

The first law of thermodynamics describes the principle of conservation of energy: in any chemical change, the total amount of energy in the universe remains constant, although the form of energy may change.

In addition, based on the well-known physics law of general relativity, explained by Albert Einstein, energy and mass are relative. Each matter (mass) even in static form contains a special level of energy which is stored in its molecules. The relativity of energy and mass is explained by the relativity equation as $E = mC^2$, where "*m*" is mass and "*C*" is light speed (approximately 3×10^8 m/s) [6].

In addition, unlike the classic physics where the light was just considered as energy in the form of the wave, Einstein asserted that light is a particle containing energy corresponding to their wavelength. Einstein succeeded to explain the photoelectric effect, which cannot be explained if one only considers light as a wave. So, light in this way came to be called a "photon (light quanta)" since it has the properties not only of a wave but also of a particle [6]. The relativity equation was initially written in many different notations. More interpretation and justification of the formula were further developed for different applications [3, 4, 7, 10]. Initially, the eq. $E = mc^2$ was not written as a formula, but Einstein described it in a sentence, as follows, "if a body gives off the energy L (in the form of radiation), its mass diminishes, which can be calculated as L/V^2" [3]. Later, he described the equivalence in the form $M = \mu + E0/ C^2$ [4]. After the Second World War, Einstein came back to the relativity equation once again and this time wrote $E = mc^2$ in the title of an article [5]. He attempted to describe the equation of relativity in a more general form.

Later, he also described the equivalence of mass and energy, clearly as follows: the energy E of physical systems numerically equal to the production of its mass (m) and the speed of light C squared. Mass-energy equivalence states that an object has a certain amount of energy, even when it is stationary. According to Newtonian mechanics, a stationary (motionless) body has no kinetic energy and it may or may not have other amounts of internally stored energy. While, based on the relativity of energy and mass, any object has an equivalent energy (E) which can be calculated as the mass (m) multiplied by the speed of light (c = about 3×10^8 m/s) squared [5, 18].

In a similar way, there are still several puzzling phenomena in nature observed in the chemical and physical properties of biological systems that cannot be explained if one just considers energy and mass as the only aspects of the matter.

This topic is further explained in Chap. 5.

Biological systems obey currently defined natural laws in physics, including thermodynamics as well as the relativity of energy and mass. However, it is not sufficient to explain all properties of biological systems. For example, autonomous properties of biological systems (including their self-organizing and self-replicating capacity) cannot be explained simply if we consider only energy and mass.

Self-organization is a remarkable property in the chemistry of biological systems that distinguish them from non-living systems in nature.

Unlike self-assembly that can be observed in both living and non-living systems (e.g., formation of ice crystals or spontaneous folding of proteins), self-organization can be observed only in living systems.

Self-assembly is a spontaneous process and as an exergonic chemical reaction does not require the uptake of energy from the environment. Therefore, self-assembly can be explained by thermodynamics and through the stability of the molecular structure. However, the formation and maintenance of a self-organizing system rely on continuous energy input into the system. Thus, by considering energy balance of the system, self-organization requires energy uptake from the environment. As another remarkable natural property, biological systems are self-fueling and can manage their energy requirements [13].

Another remarkable property of a self-organizing system is the capacity of structural symmetry breaking which also can be defined as structural polarity formation [8, 13]. Polarity formation involves autonomous generation and assembling of highly specific geometrical structures, growing through different layers of complexity (from nanostructures to micro- and macrostructures) (Figs. 1.1 and 1.9). Therefore, polarity formation in a self-organizing system requires highly specific

Fig. 1.9 (**a, b**) Comparison of self-organization and self-assembly. (**a**) Self-assembly is the process in which a disordered system with many components becomes an orderly and stable structure (at equilibrium with minimum energy configuration) without external direction. The self-assembled structure has a higher order (lower entropy) than the isolated components. Examples include the formation of ice crystals by self-assembling of water molecules at 0 °C. (Source of image: https://pixabay.com/en/ice-crystal-ice-form-frost-fabric-528528/) (**b**) Self-organization in biological systems. Image reveals highly organized structure of sunflower through multiple layers of complexity. As a distinguishing property, self-organization rely on continuous energy input into the system. Thus, by considering energy balance of the system, self-organization requires energy uptake from the environment. As another remarkable natural property, biological systems are self-fueling and can manage their energy requirements. (Source of image: https://pixabay.com/en/nature-flora-flower-leaf-sunflower-3113318/)

internal programs to guide and enforce the progress toward the formation of a desired geometrical structure. In other words, self-organization only occurs through a dedicated internally programmed process.

Even though both the self-assembly and self-organization processes refer to the spontaneous formation of orderly structure, they are different from each other in several aspects. Firstly, the self-organization is a nonlinear process. During self-organization, different types of structures can be formed at multiple layers of complexity. While molecular interaction in self-assembly is more elementary and can be traced back to the electromagnetic force. The size of the resulting structure from self-assembly is mostly mesoscopic, but it can be considerably larger for self-organization.

Considering the chemical foundation of biological systems, regulatory programs behind self-organization are required and should be integrated directly within the chemistry of the system. This involves the electrochemical interactions between molecules, which also can be defined as endogenous molecular coding.

However, the coding capacity of molecules, as well as programmability of a chemical system through the hierarchal molecular interactions, still has not been explained clearly by thermodynamics laws nor relativity of energy and mass.

The scientific explanation of self-organization in a biological system requires the consideration of the endogenous programs (codes), which are embedded in the physicochemical structure of molecules, as an additional factor beyond the energy and mass of the system.

Endogenous molecular codes, in a self-organizing system, guide the progress of process through multiple layers of complexity. The possibility of endogenous molecular coding was recapitulated from coding properties of biomolecules. Therefore, here we described this molecular coding property of biomolecules as the third dimension of nature's laws in chemistry and physics. Various aspects of this concept are fully explained in different chapters of this book.

1.6 Code: The Third Dimension of Nature's Law

The current understanding of basic natural sciences (including physics and chemistry) is focused on energy and mass as the main aspects of any matter (material) in the universe. The correlation between energy and mass as two major aspects of matter has been described clearly through the equation of relativity of energy and mass by Albert Einstein. However, still, the possibility of information storage in molecules (embedding code in molecules) remains to be described systematically as the third dimension of matter. Biomolecules such as nucleic acids are promising models for the description of the possibility of information storage in the matter (coding capacity of matter).

Here, in this book, we described the possibility of embedding information (code) in the physicochemical structure of molecules, as the third dimension of nature's law (beyond energy and mass).

Biological systems are defined as precisely programmed systems via the information storage and operation by their molecules. Nucleic acids are well-known molecules that function as highly conserved coded chemicals in nature. Deoxyribonucleic acid (DNA) is the hereditary material in humans and almost all other organisms [22]. The information in DNA is stored as a code made of four chemical bases: adenine (A), guanine (G), cytosine (C), and thymine (T). The order and sequences of these bases determine the information available for building and maintaining an organism [22]. In addition to the biological coding function of DNA, it has also been used as a matrix for storage and processing of non-biological information in a synthetic form [1, 21]. Results from this research extended to the coding capacity of DNA molecules from biology to the mathematics and computer science. For example, synthetic DNA molecules have been used as a computation matrix for solving a cluster of complex mathematical problems called nondeterministic polynomial (NP) time problems. NP problems are a class of mathematical problems with exponential complexity with no current efficient solution. Given the limitation of current silicon-based computers in sequential operation, it is almost unlikely that NP problems are solved within logical time [1]. Results of current studies on computing with DNA molecules indicate that DNA has many advantages over conventional computational methods that utilize solid-state semiconductors.

DNA molecules can execute billions of operations simultaneously and in parallel to each other. This is contrasted with the electrical digital computers where individual operations are very fast; however, the operations are executed sequentially. The massive parallelism of DNA computing comes from the huge number of molecules which chemically interact in a small volume. DNA molecules also provide a huge storage capacity in a very small space because they encode information at the molecular level. Based on the massive parallelism, DNA computing provides new solutions for NP problems.

DNA is an example of natural programmable materials. However, in addition to DNA, several other natural and synthetic materials can be applied for coding and data processing in next-generation chemical-based nanocomputers. Here, in this book, by learning from nature and considering the possibility of information storage and operation by coded molecules, as well as programmable interaction among molecules with each other and their environmental factors, we discuss three new emerging concepts in natural sciences, as follows:

1. *Programmable chemistry*
2. *Algorithmic chemistry*
3. *Cognitive Chemistry*

Programmable chemistry is a special form of chemistry that information is embedded in chemical bonds of molecules or electrochemical interactions among different molecule. In other words, molecules can be applied as a matrix for information storage in a programmable chemistry system.

Algorithmic chemistry is a special form of chemistry that information can be processed by a sequence of molecular operations. This means, in an algorithmic chemistry system, a cluster of molecules are interconnected together and can function as an algorithm. In fact, the system can function as an algorithm by a sequence of electrochemical interactions among molecules. Therefore, beyond the coding capacity of molecules, in an algorithmic chemistry system, information can be processed by a set of internally programmed molecular operators.

Cognitive chemistry is a special form of chemistry where the system is able to recognize and respond to the dynamic environmental signals and even adapt to its environmental conditions. The system is highly connected with its environmental factors through molecular sensors and logic gates. The system is made of a cluster of precisely coded molecules (that through their electrochemical interactions) are highly connected to each other as well as to their environmental factors. In fact, a cognitive chemistry system has the capacity of endogenous cognition.

In the following chapters (Chaps. 2, 3, 4, and 5), we will provide a detailed description of how cognitive chemistry can be applied in synthetic forms as an emerging tool for next-generation chemical-based computing systems. In addition, we will discuss how the application of basic principles of the relativity of code, energy, and mass can be applied to replicate the remarkable properties of biological systems including self-organization, self-regulation, decision making, etc.

1.7 Relativity of Code, Energy, and Mass: The Molecular Mechanism of Autonomy in Biological Systems

Here, by considering the possibility of information storage in molecules, we describe a new translation of relativity equation, as the relativity of code, energy, and mass.

A new translation of relativity equation explains that in nature, the matter is not limited to energy and mass, and the transformation is not limited only to energy and mass but also includes code as an additional factor.

The relativity of code, energy, and mass as a new platform in science unveils the molecular logic behind several unknown features of nature (such as natural molecular machinery and self-organization) and provides solutions for several challenges in different areas of technology and life sciences.

As an example, first we explain the trifold existence of relativity in molecules of nucleic acids (e.g., initially in molecules of adenosine triphosphate). However, it can be explained for all natural or synthetic molecules with similar multifunctional capacities of coding, energy, and mass transformation through their molecular structures (e.g., synthetic peptides, graphene, and aptamers). Given the multifunctional nature of nucleic acids, these molecules can also act as short-term carriers of chemical energy in biological systems. Above all others, the ribonucleic adenosine triphosphate (ATP) transfers energy in hundreds of different cell reactions. ATP can be formed through the recycling of energy released from the oxidative breakdown of other biomolecules. In fact, nucleic acids are molecules that nature has selected as a carrier of energy. For instance, ATP can be synthesized through the oxidative phosphorylation reactions of photosynthesis which involves the conversion of solar energy into its chemical bonds. Photosynthesis is a representative of a natural process that clearly illustrates the existence and accuracy of relativity of code, energy, and mass in nature by conversion of solar energy to chemical energy (Fig. 1.10).

Three phosphates of ATP molecule are linked in series by two phosphoanhydrate bonds, whose rupture releases large amounts of useful energy. The terminal phosphate group particularly is frequently split off by hydrolysis, often transferring a phosphate to other molecules and releasing energy that drives the energy requirements for biosynthetic reactions (Fig. 1.10). Other nucleotide derivatives (e.g., GTP, NADH) also function as energy carriers to other chemical groups.

Photosynthesis is a natural process that clearly illustrates how a programmed chemical system can generate a molecular machinery for transformation of solar energy into the mass of biomolecules. This energy also can be applied to the generation of new coded materials and self-organization of new cognitive systems.

The integration of coding capacity and the thermodynamic properties (energy and mass transformation) of biomolecules leads to the high flexibility and multifunctional properties of these molecules. In the following chapters, we will provide a detailed description of the multifunctional properties of coded molecules (e.g., in DNA and proteins) by the *relativity of code*, *energy*, and *mass*.

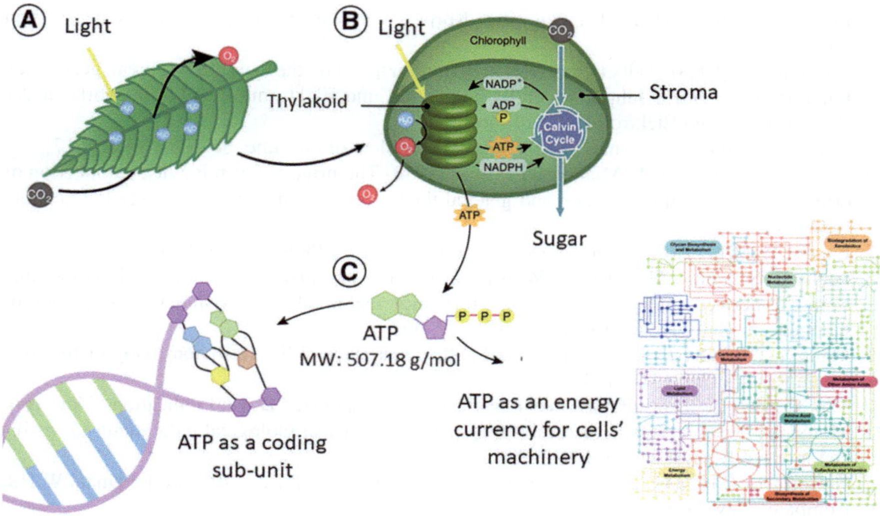

Fig. 1.10 A schematic representation of the relativity of code, energy, and mass in biological systems. (**a**) Transformation of solar energy to chemical energy (stored in ATP molecules) through photosynthesis reactions in green plants. (**b**) Transformation of solar energy to biomass (synthesis of carbohydrates) through Calvin cycle reactions in the chloroplast. (**c**) ATP as a multifunctional molecule with relativity in coding capacity, energy, and mass transformation. Unlike binary coding units in silicon-based computers which lack physical mass, DNA coding subunits (including ATP) possess a physically measurable mass (Mw of ATP = 507.18 gr)

Herein we propose that the multifunctional properties of biomolecules and the existence of relativity of code, energy, and mass in biological systems (as natural cognitive chemical systems) is the main criteria that distinguish them from other non-living, rigid materials in nature.

We will provide detailed descriptions for the molecular mechanism of self-regulation (in Chap. 2), self-organization (in Chap. 3), and self-fueling capacity (in Chap. 4) of biological systems by the relativity of code, energy, and mass.

In Chap. 6, we will discuss the application of programmable materials and *relativity of code*, *energy*, and *mass* in development of novel therapeutic approaches for the treatment of currently incurable diseases such as cancer and age-related and neurodegenerative diseases.

References

1. Adleman LM (1994) Molecular computation of solutions to combinatorial problem. Science 266(5187):1021–1024
2. Alberts B, Johnson A, Lewis J, Raff M, Roberts K, Walter P (2008) Molecular biology of the cell, 5th edn. Garland Science Taylor and Francis Group, New York

3. Einstein A (1905) Ist die Trägheit eines Körpers von seinem Energieinhalt abhängig? Ann Phys 18:639–643
4. Einstein A (1908) "Über das Relativitätsprinzip und die aus demselben gezogenen Folgerungen" (PDF). Jahrbuch der Radioaktivität und Elektronik 4:411–462 Jahrbuch der Radioaktivität und Elektronik, **4**: 411–462
5. Einstein A (1946) A, $E = mc^2$: the most urgent problem of our time. Sci Illus 1(1):16–17
6. Einstein A, Lorentz HA, Minkowski H et al (1952) The principle of relativity, a collection of original memoirs on the special and general theory of relativity, 1st edn. Cover Publication, New York
7. Eugene E (2011) How Einstein confirmed E0=mc2. Am J Phys 79(6):591–600
8. Fuchs C, Scheinast M, Pasteiner W et al (2012) Self-organization phenomena in embryonic stem cell-derived embryoid bodies: axis formation and breaking of symmetry during cardiomyogenesis. Cells Tissues Organs 195(5):377–391
9. Mann S (2012) Systems of creation (2012) the emergence of life from non-living matter. Acc Chem Res 45(12):2131–2144
10. Max J (1997) Concepts of mass in classical and modern physics. Dover Publications, New York
11. Moreno A, Etxeberria A, Umerez J (2007) The autonomy of biological individuals and artificial models. Biosystems 91:309–319
12. Nelson DL, Cox MM (2017) Lehninger principles of biochemistry, 7th edn. Freeman, W H& Company, New York
13. Northrop B, Zheng Y, Chi K, Stang P (2009) Self-organization in coordination-driven self-assembly. Acc Chem Res 42(10):1554–1563
14. Prescher JA, Bertozzi CR (2005) Chemistry in living systems. Nat Chem Biol 1(1):13–21
15. Reece JB, Urry LA, Cain SA, Wasserman PV, Minorsky PV, Jackson RB (2011) Some properties of life. In: Campbell biology, 10th edn. Pearson, San Francisco
16. Rodwell VW, Bender D, Botham K, Kennelly P, Weil PA (2015) Harpers illustrated biochemistry, 30th edn. The McGraw Hill Education, US, New York/London
17. Sadava DE, Heller HC, Hillis DM, Beren-Baum MR (2009) What is life? In: Life: the science of biology, 9th edn. Sinauer Associates, Sunderland
18. Schwartz HM (1977) Einstein's comprehensive 1907 essay on relativity, part II. Am J Phys 45(9):811–817
19. Shahbazi MN, Jedrusik A, Vuoristo S et al (2016) Self-organization of the human embryo in the absence of maternal tissues. Nat Cell Biol 18(6):700–708
20. Steiner J (2006) The origin of universe. Estudo Avancados 20(58):233–248
21. Thubagere AJ, Thachuk C, Berleant J et al (2017) Compiler-aided systematic construction of large-scale DNA strand displacement circuits using unpurified components. Nat Commun 8(14375):1–12
22. Watson JD, Crick FC (1953) Molecular structure of nucleic acids. Nature 171(4356):737–738

Chapter 2
Molecular Mechanism of Coding and Autonomous Decision-Making in Biological Systems

Abstract Biological systems are recognizable from inanimate materials through their cognition and computation capacity. Cells are the main subunits of a biological system and function as highly advanced computers by executing thousands of operations per second for different biological purposes to dynamically adapt with the environment. Unlike current electronic-based computers, biological systems utilize a molecular-based coding system in which information is stored in molecules. Information storage in molecules provides massive operation capacity for the cells. Deep understanding of mechanisms of coding and data processing in the cells could have several technology applications and trigger an industrial revolution. However, this level of progress requires the establishment of a different scientific viewpoint for life sciences – a paradigm that puts life sciences in a category that is much closer to the other experimental branches of natural sciences including chemistry, physics, and mathematics.

In this chapter, first we provide a detailed description of different aspects of molecular coding and data operation in biological systems applying new concepts of *cognitive chemistry* and the *relativity of code*, *energy*, and *mass*. We will discuss how information is stored in the patterns of molecular interactions and how real-time interactions between molecules and atoms generate a dynamic coding and operation capacity in biological systems. In the second part, we will discuss how we can leverage the cognitive chemistry knowledge in designing synthetic systems with similar autonomous properties of biological systems. In the third part of this chapter, we will discuss how basic principles of cognitive chemistry can be applied to mimic the extensive computation capacity of biomolecules in solving complex decision making problem.

Keywords Cognitive chemistry coding system · Nondeterministic polynomial time problems · Molecular computing · Artificial Intelligence (AI) · DNA computing · Multilayer coding · Conserved and dynamic coding · Stem cells and decision-making · Eternal cognition

T. Karimi, *Molecular Mechanisms of Autonomy in Biological Systems*,
https://doi.org/10.1007/978-3-319-91824-2_2

2.1 Molecular Mechanism of Autonomous Coding and Data Processing Based on Physical and Chemical Foundation of Life

Considering the physical and chemical foundations of life, it can be concluded that living organisms are physicochemical systems and are not exempt from nature's laws in physics and chemistry. Scientifically, we know that biological systems are composed of normal elements of nature. However, the regulatory mechanisms that for the first time caused the conversion of elements from randomness into highly organized system (determined as life) have remained a mystery.

"Science cannot solve the unlimited mystery of nature. And that's because, in the last analysis, we ourselves are part of nature and therefore, part of the mystery that we are trying to solve". (Max Planck).

The precise regulatory mechanisms behind "the origin of life" five billions years ago when the first self-generating reaction of life occurred remained a nature's mystery (due to our limitations in the experience of time). However, the silver lining is that we can learn from the currently existing rules of life that already are accessible to human experience.

In the early twentieth century, right after the establishment of modern physics and the revolutionary effect of quantum physics on different technologies, Erwin Schrodinger referred to the emergence of a revision in the definition of life. Furthermore, he attempted to describe the revision in the classification of natural sciences by filling the gaps between life and other branches of sciences. In his well-known book about life, titled "What Is Life," he referred to the missing parts in the classification of different branches of natural sciences that separate living systems from other human-made physicochemical systems. He described this gap to be related to our limited understanding of the structure and construction of living systems. Said another way, the construction of living systems is different from anything we have yet tested in the human-made physical laboratories.

"What I wish to make clear in this last chapter is, in short, that from all we have learned about the structure of living matter, we must be prepared to find it working in a manner that cannot be reduced to the ordinary laws of physics. That not on the ground that there is any new 'force' or what not, directing the behavior of the single atoms within a living organism, but because the construction is different from anything we have yet tested in the physical laboratory" [25].

During the early twentieth century, along with the great progress in physics and chemistry and later in mathematics (data science), a revolutionary movement occurred in conventional classifications of natural sciences (Fig. 2.1a, b). Different branches of natural science including physics, chemistry, and mathematics moved closer to each other, and subsequently, several new fields were established in natural sciences (Fig. 2.1b). For example, modern (quantum) physics was established by a new interpretation of integrated physics and chemistry. Later, computer science was established by integration of electronic physics and data science (mathematics) (Fig. 2.1c). During the second world war, Alan Turing developed the first universal digital computer (also called Turing machine). He worked on algebra and number theory, as well as a cipher machine based on electromagnetic relays to multiply binary numbers.

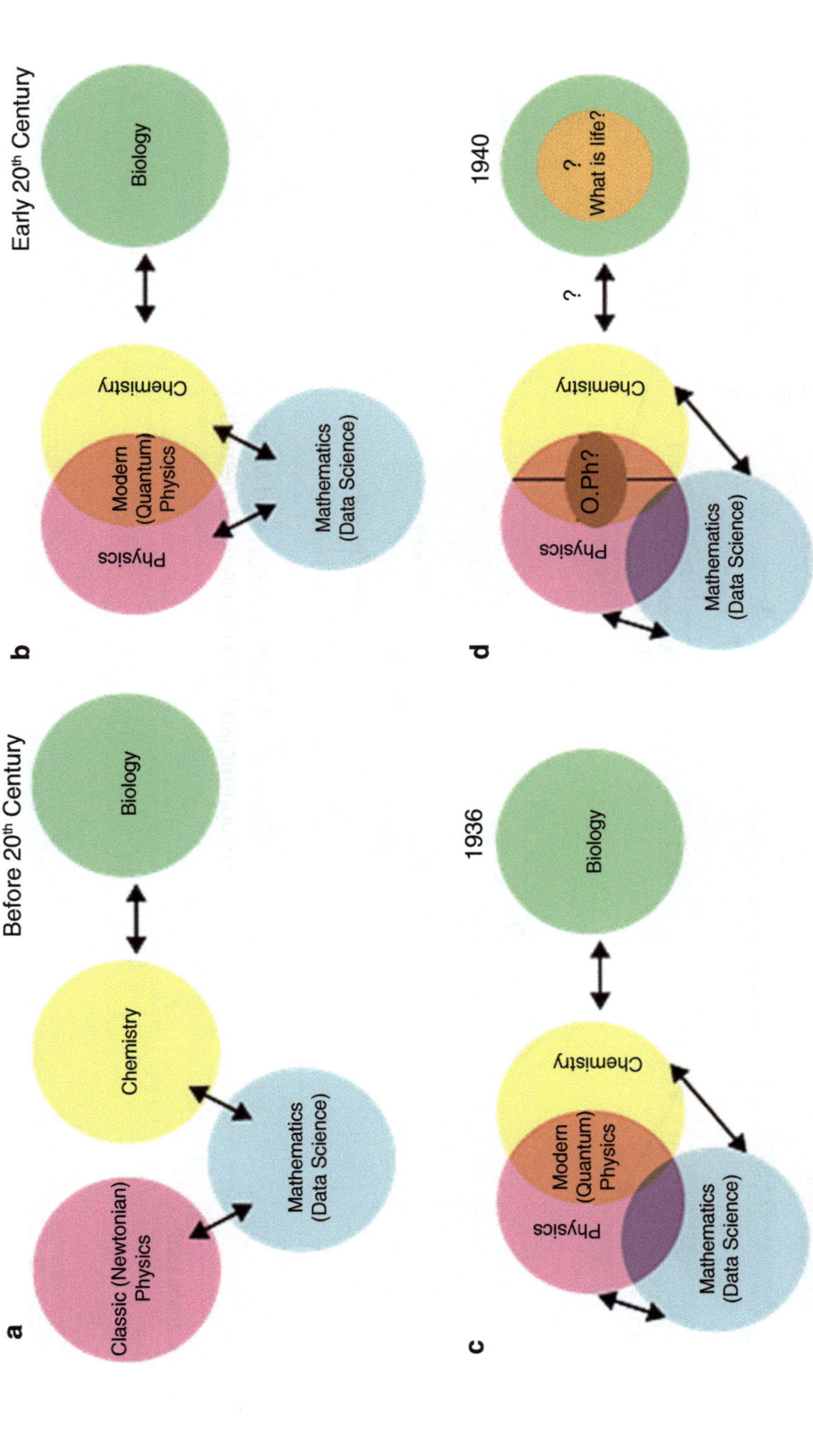

Fig. 2.1 **(a–g)** Schematic image illustrating the trend of changes in the classification of natural sciences after the twentieth century as a result of the cross-pollination or integration of basic natural science. TC, Turing Computing; QC, Quantum Computing; O.Ph, Ordered Physics

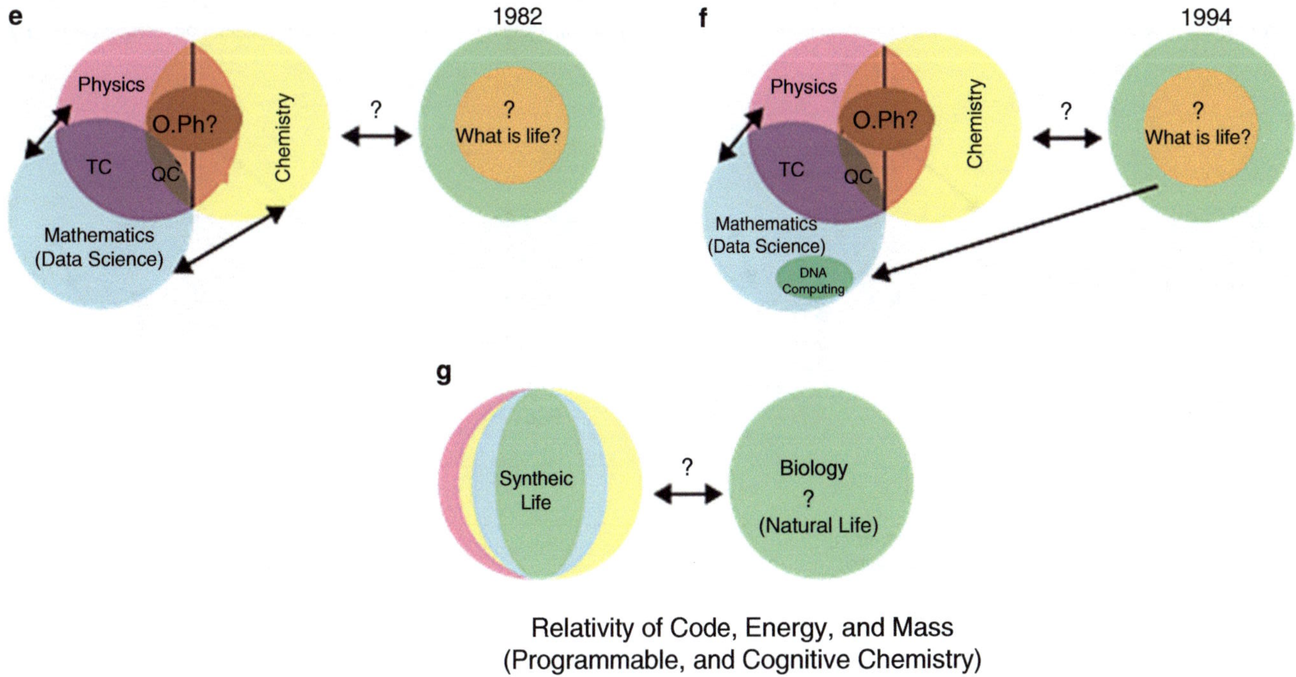

Fig. 2.1 (continued)

At the early twentieth century, basic principles of modern physics were established by Max Plank. Modern physics later became highly practical and were applied in different technologies by research work of Albert Einstein and Erwin Schrodinger on quantum mechanics.

At 1900, Max Planck tried to explain blackbody (BB) radiation from warm objectives. Then, the BB spectrum was quantified by assuming that the energy in an electromagnetic wave, according to Max Planck eq. ($E_{\text{EM Wave}}$ = nhf).

At 1905, Albert Einstein explained the concept of a photon as a pocket of electromagnetic radiation which behaves as energy (wave) and particle (mass) at the same time. Also, he described the equation for the relativity of energy and mass as $E = MC^2$ [3].

Later at 1927, Erwin Schrodinger attempted to define a wave equation that could precisely describe the energy in an atom particle. Schrodinger succeeded to present an equation that could precisely predict energies in hydrogen atoms which led to the establishment of the basic principles of quantum physics. Schrodinger's equation probably was the most important equation of the twentieth century, because of its huge effect on technological progress. The major significance of Schrodinger's work in quantum physics comes from its broad coverage of both physics and chemistry.

The physicist, Paul Dirac famously asserted: "*The Schrodinger equation accounts for much of physics and all of chemistry*".

In addition, at the mid-twentieth century, Schrodinger attempted to revise the classical definition of life sciences by application of basic principles of physics and chemistry in biological systems. The full manuscript of his lecture series about the physicochemical foundation of life was later published as a book titled "What Is Life." [25]

Schrodinger attempted to provide several pieces of evidence to prove the accuracy of principles of quantum physics in biological systems. He believed differences between biological systems and other physical and chemical systems are simply related to their structures. However, still, they are made of natural elements and obeying all nature's laws in physics and chemistry.

He tried to explain that differences in the construction of two systems may cause a big difference in their function, while still the basic elements applied in both systems might be the same. This needs to be differentiated from anything else that may cause biased understanding of basic elements and nature of a system. The difference in construction is enough to make an entirely different way of functioning. Schrodinger tried to make this concept clear by comparing the structural differences between a heat motor and an electronic circuit, both of which may apply the same basic elements (e.g., copper and iron). He referred to the consistency of the natural properties of copper even though it has been applied differently in the construction of a heat motor compared to an electronic circuit.

"To put it crudely, an engineer, familiar with heat engines only, will after inspecting the construction of an electronic motor, be prepared to find it working along principles which he does not yet understand. He finds the copper familiar to him in kettles used here in the form of long, wires wound in coil; the iron familiar to him in levers and bars and steam cylinders here filling the interior of those coils of copper wire. He will be convinced that it is the same copper and the same iron, subject to the same laws of nature, and he is right in that. The difference in construction is enough to prepare him for an entirely different way of functioning [25]."

The concept of consistency in the subject of nature's law earlier has also been mentioned by Max Planck and Einstein. However, Schrodinger attempted to provide a systematic description for the consistency of nature's laws in living systems aligned with quantum mechanics.

All matter originates and exists only by virtue of a force which brings the particle of an atom to vibration and holds the most minute solar system of atoms together.

We must assume behind this force the existence of a conscious and intelligent mind. The mind in the matrix of all matter.

Max Planck

Reflections of this type made it clear to me as long ago as shortly after 1900, i.e., shortly after Planck's trailblazing work, that neither mechanics nor electrodynamics could (expect in limiting cased) claim exact validity.

Gradually I despaired of the possibility of discovering the true laws by means of constructive efforts based on known facts. The longer and the more desperately I tried, the more I came to the conviction that only the discovery of a universal formal principle could lead us to assured results... How, then, could such a universal principle be found?

Albert Einstein

Schrodinger referred to the "orderliness" as the major difference between a living system and anything else that exists in inanimate matter.

"The unfolding of events in the life cycle of an organism exhibits an admirable regularity and orderliness, unrivaled by anything we meet with in inanimate matter. We find it controlled by a supremely well-ordered group of atoms, which represents only a very small fraction of the sum total in every cell."

As examples of ordered structures in nature, Fig. 2.2 illustrates the interconnections of living systems with inanimate elements in nature, and Fig. 2.3 illustrates the spontaneous formation of a highly organized structure in a snail shell from disordered elements of nature which is defined as ordered physics by Erwin Schrodinger.

In addition, based on the behavior of atoms, Schrodinger classified all physical systems in nature into two main categories including *statistical physics* and *order from order* (ordered) physics (Fig. 2.1d).

"I remember an interesting little paper by Max Planck on we have the topic the dynamical and statistical type of law (Dynamische und Statistische Gresetzmassigkeit). The distinction is precisely the one we have here labeled as

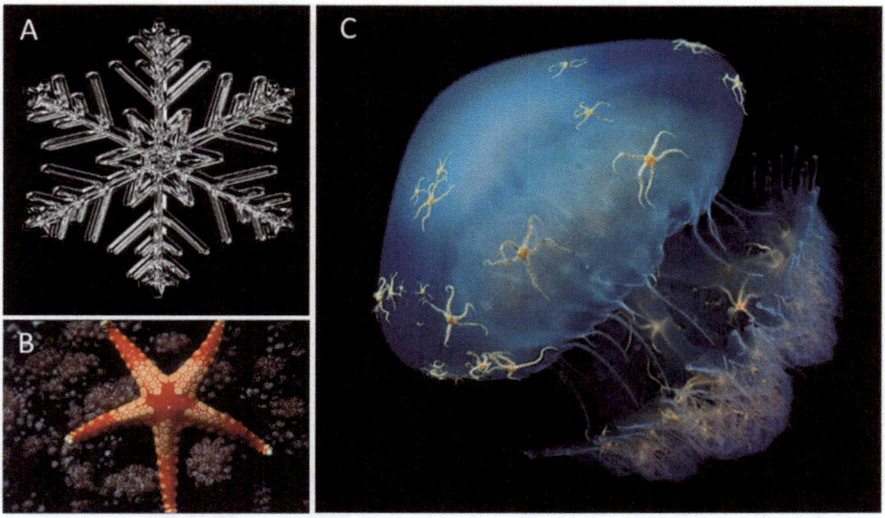

Fig. 2.2 (**a-c**) provides examples of spontaneous formation of highly organized structures in nature among non-living and living systems. A- Formation of ice crystal (non-living). B- Geometrical structure of a starfish C- A jellyfish. Living systems are made of natural elements of nature and are obeying all nature's laws in physics and chemistry. What differentiates the chemistry of a living system from non-living elements of nature is related to the capacity of performing autonomous functions (e.g self-regeneration, self-regulation and self-fueling capacity) applying their molecular units

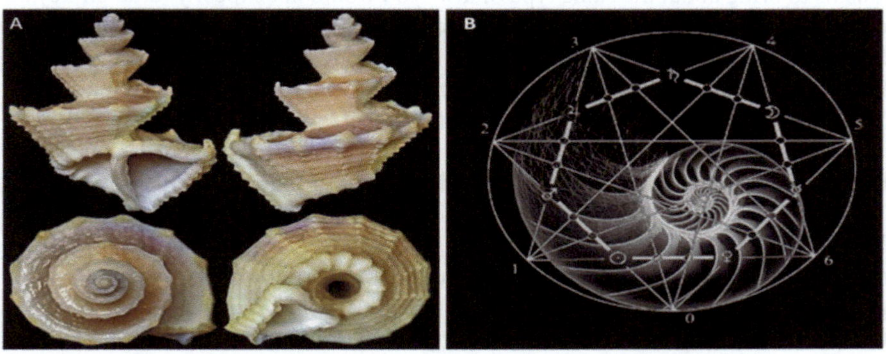

Fig. 2.3 (**a, b**) Autonomous formation of highly organized geometrical structures in a snail shell. Formation of highly organized structure from disordered elements of nature is called ordered physics by Erwin Schrodinger

'order from order' and 'order from disorder' [25]. The object of that paper was to show how the interesting statistical type of law, controlling large-scale events, in constituted from the dynamical laws supposed to govern the small-scale events, the interaction of the single atoms and molecules. The latter type is illustrated by large-scale mechanical phenomena, as the motion of the planets or of a clock, etc. thus it would appear that the new principle, the order- from- order principle, to which we have pointed with great solemnity as being the real clue to the understanding of life."

What is life?

Is life based on the laws of physics?

From all we have learnt about the structure of living matter, we must be prepared to find it working in a manner that cannot be reduced to the ordinary laws of physics.

And that not on the ground that there is any "new force" or what not, directing the behavior of the single atoms within a living organism, but because the construction is different from anything we have yet tested in the physical laboratory.

The unfolding of events in the life cycle of an organism exhibits an admirable regularity and orderliness. We must be prepared to find a new type of physical law prevailing in it.

Erwin Schrodinger

"For the new principle that is a genuinely physical one, it is, in my opinion, nothing else than the principle of quantum theory over and over again. But, we cannot expect the 'laws of physics' derived from it suffice straightaway to explain the behavior of living matter whose striking features are visibly based to large extent on the 'order-from-order' principle. You would not expect two entirely different mechanisms to bring about the same type of law- you would not expect your latch-key, to open your neighbor door as well. We must therefore not be discouraged by the difficulty of interpreting life by the ordinary laws of physics. For that is just what is to be expected from the knowledge we have gained of the structure of living matter. We must be prepared to find a new type of physical laws prevailing in it." [25]

Now the big question is what would be the new type of physical laws that prevail the highly ordered structure, function, and behavior of living systems?

Here in this book, we attempt to explain the mechanisms that cause the *ordered behavior* of molecules and atoms in biological systems. We will discuss how *order from order physics* can be explained by unique property of biological systems for generation of molecular codes. As an additional factor in nature's laws, we describe the coding capacity of molecules in biological systems that should be considered alongside the properties of molecules for energy and mass production. Therefore, we determined code as the third dimension of nature's laws that have been illustrated in the autonomous properties of biological systems.

By considering code as the third dimension of nature's laws in physics and chemistry, we determined a new branch of chemistry called *cognitive chemistry* that mimics the *ordered physics* in biological systems. Cognitive chemistry integrates nature's laws in chemistry, physics, and mathematics for coding, data storage, and operation by molecules (Fig. 2.1g). Cognitive chemistry can be interpreted in physics terminology as the *relativity of code*, *energy*, and *mass*.

2.2 Basic Principles of Coding and Data Processing in Biological Systems

2.2.1 Information Storage in Molecules and Materials

Unlike human-made computers that are operated according to physical- and electrical-based coding, biological systems apply a unique chemical-based coding system. In this system biological information is embedded in materials or chemical interactions between molecules and atoms. For simplicity, in this book, code-embedded materials are called "coded materials". As an example, deoxyribonucleic acid (DNA) is the most well-known coded material that nature utilizes for storage of biological information. DNA, also known as genetic material, contains the entire information of an organism to be copied into the next generation of the species [17]. DNA provides a huge storage capacity because DNA encodes data applying four subunits including A, G, C, and T, while current computers apply a binary (0,1) coding system. Neurotransmitters are other examples of coded materials which play a major role in the information storage and transduction in the neural system.

Figure 2.4 compares different features of information storage and data processing in the molecular-based coding system of a seed (A), versus the binary coding system in an electronic-based flash memory (B). In fact, in a molecular coding system, the software (coding unit) and hardware (structural unit) are integrated. For instance, in a DNA coding system, each coding unit (AGCT) possesses a physical mass. While in an electronic coding system, each coding units (0 or 1) lacks a molecular mass and is generated just by electronic currents. In addition, self-organization and autonomous

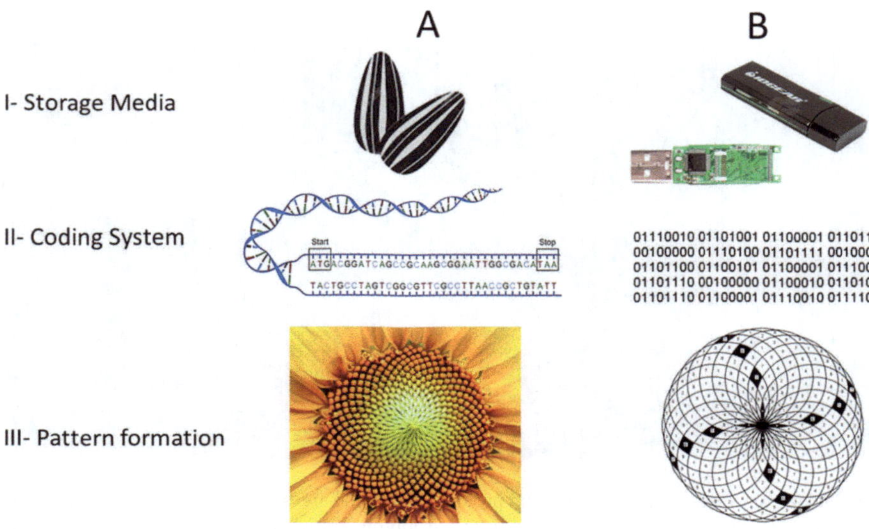

Fig. 2.4 **(a–b)** Illustrates examples of chemical-based coding and information storage in plants' seeds versus electronic-based coding and information storage in an electronic memory applying binary coding system

formation of structural patterns are the direct results of data operation in a molecular coding system. For example, all information storage for the synthesis of highly organized geometrical structures of a sunflower has been saved in a physical mass of a seed.

Figures 2.5, 2.6, 2.7, and 2.8 provide examples of chemical-based coding in the neural system. The entire brain mass is generated from coded material. In fact, the brain is a chemical-based computer which works based on fundamental principles of cognitive chemistry rather than statistical chemistry (or molecular randomness and statistical possibility in Schrodinger's definition for a non-living system).

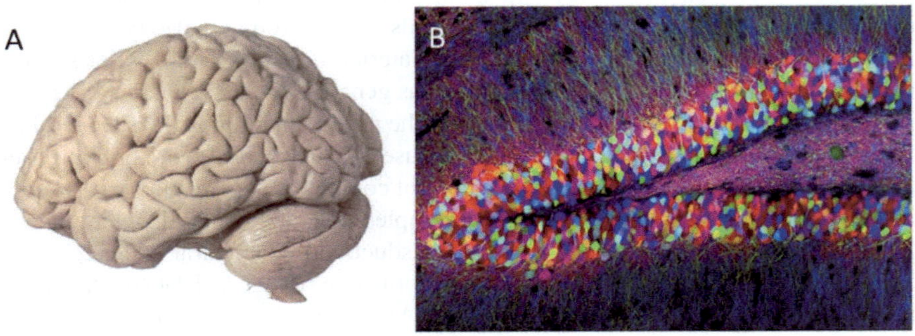

Fig. 2.5 (**a**, **b**) The macroscopic and microscopic structure of the brain. The physical mass of the brain, which is totally made of coded molecules, provides a highly efficient matrix for information storage and memory formation. (**b**) Microscopic imaging of mice brain applying Brainbow method and mosaic expression of fluorescent proteins in neural cells. Microsections of mice brain were taken from dentate gyrus of the hippocampus (Image in part **b** is adapted from Weissman, et al., [36])

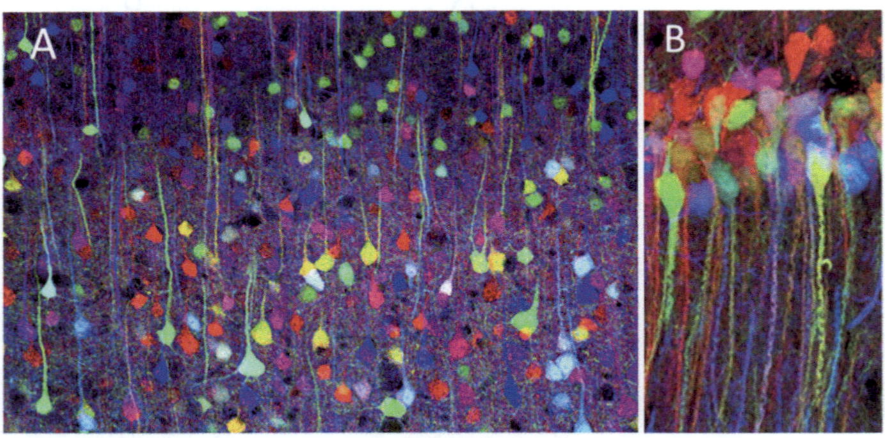

Fig. 2.6 (**a**, **b**) Coexpression of florescent proteins in specific areas of brain by Brainbow method. (**a**) Hippocampus neurons. (**b**) Cortex neurons. Applying Brainbow method, neural cells were labeled with FPs that specifically are coexpressed with molecular markers for different parts of the brain. Image obtained at higher magnification reveals the whole neural networks of brain are wired by information-enriched molecules (more than electrical neural networks) (images are adapted from [36])

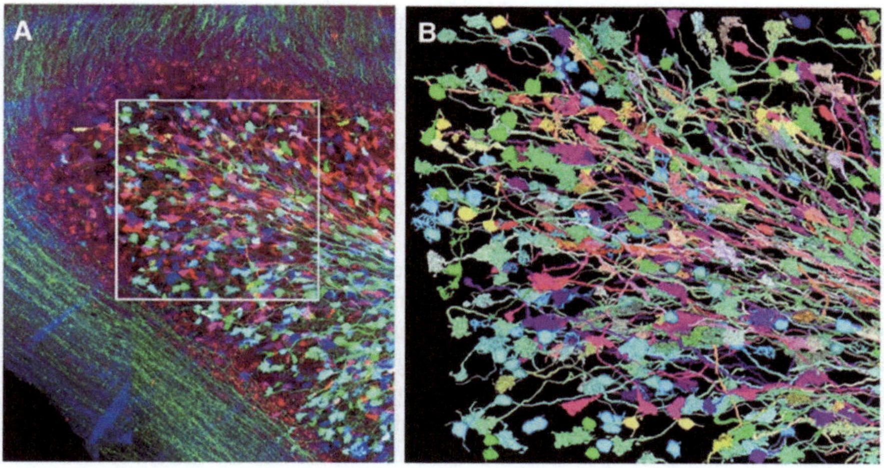

Fig. 2.7 (**a**, **b**) Microscopic images of neural cells, illustrating the direct role of molecules of neurotransmitters in both data transportation as well as memory formation in neural synapses and consequently neural networks. (**a**) A cerebellar folium from the Brainbow mouse line H was imaged using confocal microscopy. Three-dimensional volume indicated in the box was segmented using semiautomated methods and reconstructed digitally. (**b**) Digital reconstruction of axons and granule neurons from volume marked in A (Images are adapted from [34])

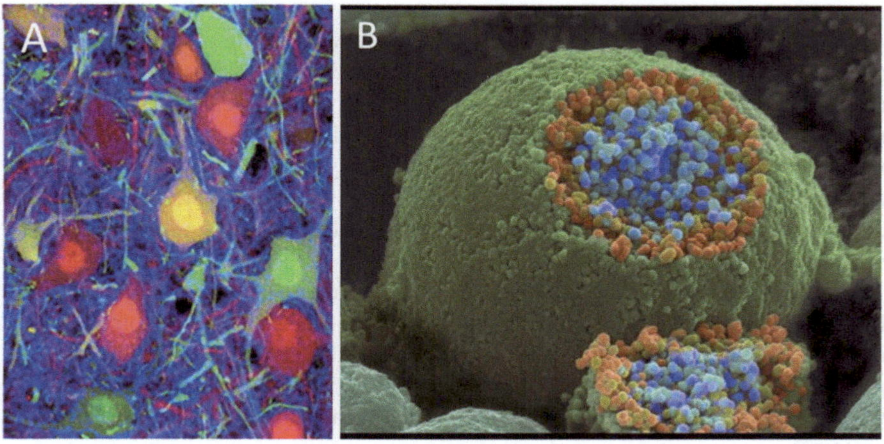

Fig. 2.8 (**a–b**) Microscopic images of labeled neural cells at high magnification by dark-field microscopy as well as electron microscopy. (**a**) Microscopic imaging of neurons by expression of different FPs with various neurotransmitters. (**b**) A colorized scanning electron microscope image of a nerve ending that has been broken open to illustrate the synaptic vesicles containing information-enriched molecules of neurotransmitters (orange and blue) beneath the cell membrane. Unlike the electronic coding system in silicon-based computers, the major mass of neurons is generated from information-enriched molecules. In fact, a unique property of brain's coding system compared to electronic coding system is the integration of hardware and software in the whole physical mass of the brain. (Source of images: **a** – [36]. **b** – http://www.cellimagelibrary.org/images/214)

Figure 2.5a illustrates the physical mass of brain, which is the initial origin of memory formation. The brain is the only computing system that has the capacity to generate expandable memory volume in its limited physical mass. The main difference between the brain and current electronic computers is in the brain's molecular-based coding system. This molecular coding system is the origin of electrical activities of neural cells. Furthermore, memory formation originates from the molecular coding system of neural cells. While electrical coding in silicon-based computers is based on transportation of electrons over a uniform metal (e.g., in wires and transistors) media, the brain is made of soft material.

Figure 2.5b illustrates an immunohistochemistry (IHC) imaging of brain microsections. Tissue microsections are stained against specific molecular markers of neural cells at different layers of the brain. Figure 2.6a–e illustrates IHC staining of the brain. Microsections were stained against molecular markers of various types of brain neural cells at different layers of the brain.

Figure 2.7 illustrates the direct role of the molecules of neurotransmitters in both data transportation as well as memory formation through their role in the formation of neural synapses and consequently neural networks. Neurotransmitters are a class of coded chemical molecules that play a major role in the formation of memory through neural networks. Neurotransmitters function as chemical (molecular)-based logic gates and switches. Initiation of the action potential and electrical activity of neurons are regulated by the logic gate function of neurotransmitters at the location of neural synapses. Consequently neural networks are forming by coupling of neural synapses together.

Figure 2.8a illustrates microscopic imaging of IHC staining of the brain sections against different types of neurotransmitters in neural cells. Microscopic imaging of neural cells at high magnification illustrates the excretion of neurotransmitters from neural cells into synaptic spaces. Unlike binary coding in silicon-based computers which are completely an electronic-based coding system, data operation in neural cells involves both molecular and electrical coding systems. In fact, even the electrical activity of neural cells is originated from molecular coding by neurotransmitters, which possess a specific molecular mass.

Figure 2.8b illustrates electron microscopy imaging of a neural network. Figure 2.8c illustrates electron microscopy imaging of vesicles of neurotransmitters in a neuron. Unlike an electronic coding system in silicon-based computers that lack physical mass, the major mass of neurons as coding subunits of a neural network is generated from coded materials (e.g., vesicles of neurotransmitters and signaling proteins).

Considering that basic difference of brain operation system compared to electronic computers is related to the existence of molecular-based coding system (through different types of coded materials), it can be concluded that unique capacities of the brain (e.g., expandable memory, learning, etc.) can be explained and even simulated in synthetic form through innovative molecular-based coding systems for next-generation cognitive systems.

2.2.2 *Multilayer Coding*

Unlike current silicon-based computing systems that utilize one layer of linear binary coding system, biological systems apply multiple layers of coding. In addition, each layer of coding in a biological system involves different types of coding subunits. For example, multilayer coding in biological systems involves coding languages of DNA, mRNA, amino acids, peptides, protein-protein interaction, signaling pathways, systemic signaling pathways of endocrine hormones, as well as neurotransmitters and neural networks. Figure 2.9a illustrates schematic images representative of DNA, amino acid, and protein-coding layers in a cell versus linear binary coding in an electronic-based computer. Figure 2.9c illustrates schematic representative of standard DNA-amino acid codon in biological systems.

Despite several attempts for generation of biologically inspired computing algorithms, the current biologically inspired algorithms almost are established only based on one layer of coding. Genetic algorithm, ant colony, and DNA computing are the most popular biologically inspired computing algorithms [5, 6, 10, 21, 28, 33].

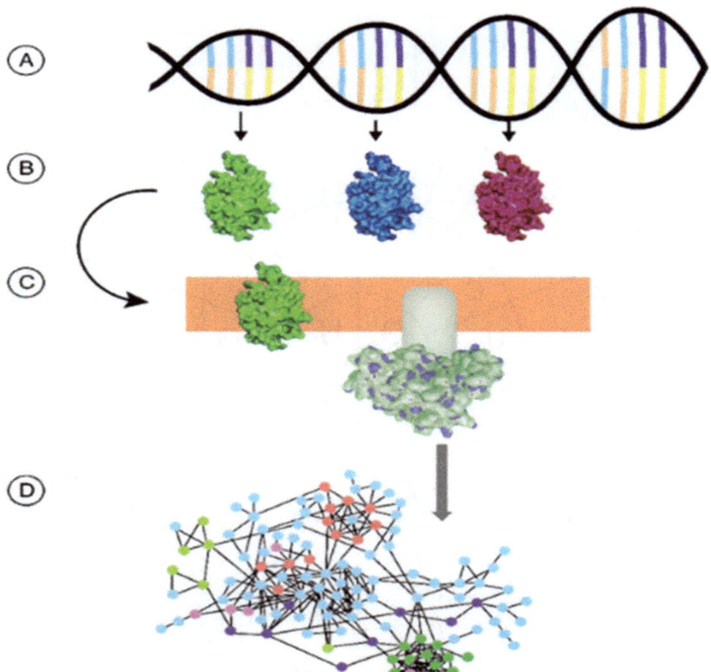

Fig. 2.9 (**a–d**) Schematic representative of multilayer coding in biological systems. (**a**) Formation of genetic coding layer by special sequences of nucleotides in DNA molecule. (**b**) Translation of DNA genetic code from nucleotides' coding language to amino acids' coding system and formation of a protein which would be a coding subunit of a cell signaling network. (**c**) Interactions of signaling protein units with each other and formation of a protein-based cell signal transduction network. (**d**) Dynamic data operation in a cell by protein-based cell signaling networks

The exquisite accuracy and efficiency of data operation in living cells through the highly complex and interconnected DNA-protein NP networks motivated us to determine a computing algorithm which is leveraged from the wide computation potentials of both nucleic acids and proteins for solving non-biological NP problems (which would be discussed in the following parts of this chapter). Figure 2.10 illustrates a schematic image representative of multilayer property of the biological coding system. Codes are defined by alphabetic symbols. The nucleic acid coding layer is made of four coding subunits and protein-coding layer is made of 20 amino acid coding subunits.

2.2.3 Molecular Coding and Algorithmic Chemistry

Unlike a silicon-based computers that is operated by an electronic-based coding system, all biological functions in living cells are operated by chemical-based coding systems. Figure 2.10 illustrates alphabetic symbols of the universal nucleotide/amino acid coding system in living systems.

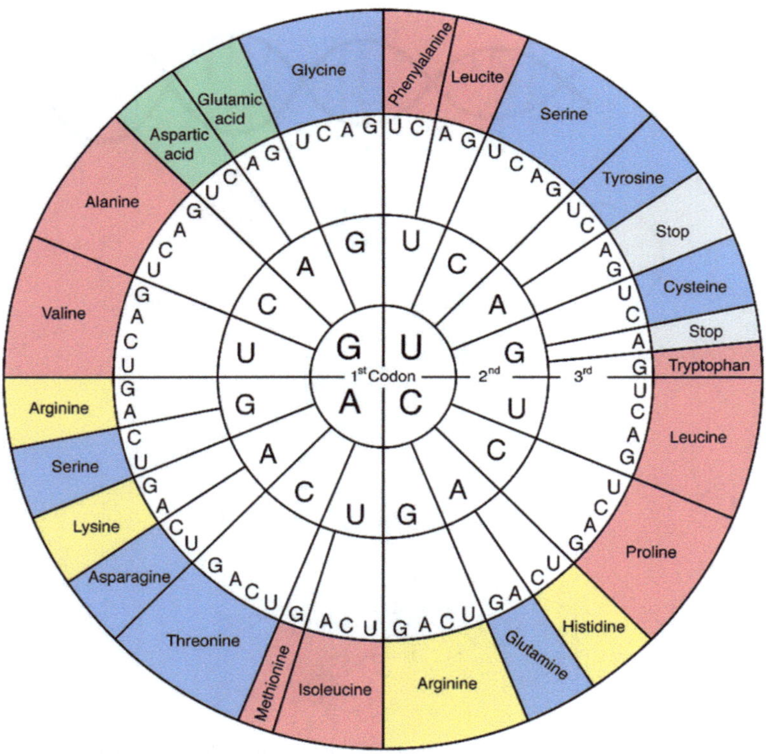

Fig. 2.10 Schematic image representative of alphabetic symbols of nucleotide and amino acids in DNA, RNA, and protein-coding layers, respectively. The nucleic acid coding layer is made of four coding subunits including A, U, G, and C. Protein-coding layer is made of 20 amino acid coding subunits

Although typically molecular codes in cognitive chemistry system are presented as sequences of molecular subunits (e.g., with sequences of 4 DNA nucleotides or 20 amino acids), the real codes in this molecular coding system are hidden in electrochemical attractive forces among molecules and atoms. Figure 2.11a illustrates a schematic image representative of electrochemical attractive forces among molecules which can be applied in designing of different types of coding molecules. Figure 2.11b indicates a schematic image representative of electrochemical interactions between two amino acids (aspartate and lysine). Dynamic electrochemical interactions between amino acids in a protein or among different proteins provide a highly efficient coding and signal transduction system in living cells. Figure 2.11c indicates electrochemical interactions between nucleotides A-T and G-C.

Electrochemical attractive forces among coding molecules can be ordered in the form of various molecular algorithms for different computational purposes. For example, electrochemical forces among coding molecules can be translated into values and integers based on the level of attraction strength between chemical bonds.

The significance and capacities of cognitive chemistry coding and data operation system include:

- The massive information storage and data compaction capacity through multiple layers of coding.
- The capacity of massive parallel operation applying numerous molecular operators.
- The autonomous data operation and environmental recognition capacity through the electrochemical interactions of coding molecules with each other, as well as the physicochemical environmental factors. Autonomous data operation and

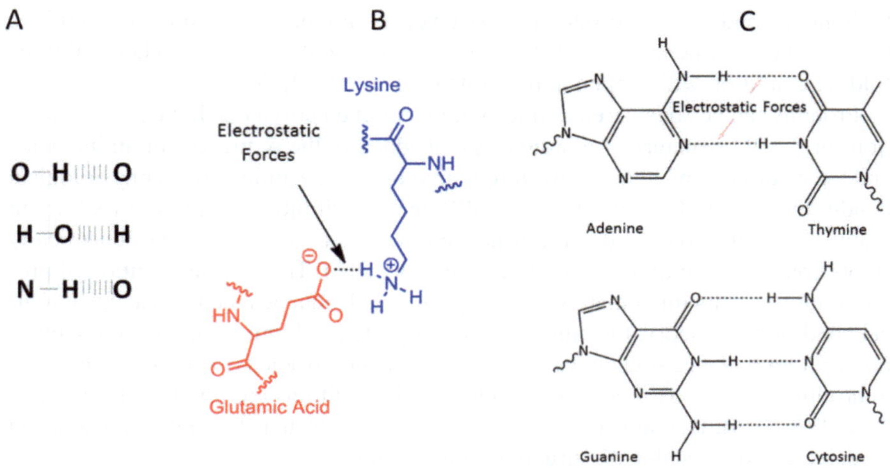

Fig. 2.11 **(a–c)** Schematic image representative of electrochemical interactions between four coding subunits of DNA nucleotides as well as the side chains of two exemplary amino acids. (**a**) Illustrates a schematic image representative of electrochemical attractive forces among molecules which can be applied in designing of different types of coding molecules. (**b**) Indicates a schematic image representative of electrochemical interactions between two amino acids (aspartate and lysine). (**c**) Indicates electrochemical interactions between nucleotides A-T and G-C

environmental recognition capacity of coding molecules can be explained by quantum mechanics effect of atoms on each other (discussed in Chap. 3).

- The spontaneous and combinatorial nature of molecular interactions, at each coding and operation layer, leads to the real-time and massive operation capacity of a cognitive chemistry coding system (e.g., for solving complex mathematical problems).
- Electrochemical attractive forces among different types of coding molecules can be quantified by specific affinity indexes. Affinity indexes among coding molecules can be translated to values and integers and be applied in designing of novel chemical-based operating systems for solving complex mathematical problems such as nondeterministic polynomial time (NP) problems. Biological systems provide highly efficient models for solving complex problems. In the following part of this chapter, we will discuss the application of biomimetic approaches for solving NP problems.

Figure 2.12 illustrates a schematic images representative of a protein folding-related NP problem. Figure 2.12.I illustrates a schematic image representation of a sequence of amino acids in an exemplary protein before folding. Figure 2.12.II illustrate schematic image representation of the electrochemical attractive forces. Amino acids are determined as nodes and electrochemical interactions between amino acids are determined by edges. The schematic image illustrates different possibilities of protein folding through all various combinatorial electrochemical attractive forces among amino acids. As it can be observed in the picture, the combinatorial nature of the molecular interactions makes a NP network for the protein folding problem. Therefore, for prediction of 3D structure of proteins applying current silicon-based computers, by increasing the number of amino acids, the operation time increases exponentially. However, protein folding is a highly precise process in biological conditions. Despite the existence of all protein folding possibilities through the electrochemical interactions of amino acids with each other, only one folding condition will happen in real time (Fig. 2. 12. III, IV).

Proteins can be considered as information enriched polymers. In fact, the instruction for the 3D structure of a protein is embedded in the sequences of amino acids and the electrochemical attractive forces among them. Within a given physiological condition, despite the existence of all different possibilities for protein folding, in reality, only one folding condition happens which is the most electrochemically stable conformation of a protein molecule (Fig. 2.12). The optimal solution of protein folding occurs immediately, after exposure with aqueous media in a physiological condition. In theory, the high accuracy of protein folding from the viewpoint of possibilities and electrochemical forces between amino acids is similar to the Erwin Schrodinger's cat box theory which means all conditions are possible at the same time (Fig. 2.13). But, at each condition only one molecular folding forms and can be detected with a stable 3D structure, immediately.

The optimal solution of a protein folding problem is highly accurate and repeatable because it provides the most stable spatial conformation through the Gibbs free energy level for the molecule. Protein folding mechanism can be simulated for solving nondeterministic polynomial time (NP) problems.

The electrochemical properties of amino acids can be applied as the coding and operational criteria for several computable functions, for example, designing of soft

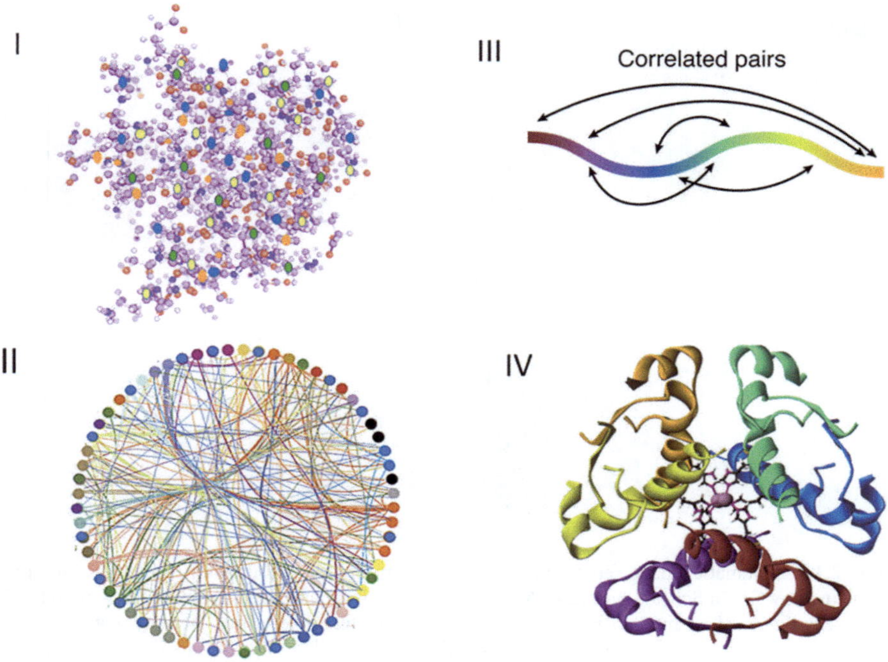

Fig. 2.12 Schematic images representative of a protein folding-related NP problem and the visualization of various possible scenarios of folding through electrochemical attractive forces among amino acids. (**I**) Schematic image representative of the sequence of amino acids in a protein before folding. (**II**) Schematic image representative of visualization of electrochemical interactions among amino acid sequences in an exemplary protein. Amino acids are determined as nodes and electrochemical interactions among amino acids are determined by edges. The schematic image illustrates different possibilities of protein folding through the electrochemical interaction of amino acids with each other. (**III**) Despite the existence of all different protein folding possibilities through the electrochemical interactions of amino acids' side chains with each other, only one folding condition will happen in real time. (**IV**) The optimal solution in a complex NP problem of protein folding is the most electrochemically stable conformation of protein molecule and can be detected by the crystallography of protein. The optimal solution of the problem is highly accurate and repeatable because it provides the most stable molecular structure through the Gibbs energy level for the molecule

logic circuit networks applying the logic gate function and signal-transducing activities of proteins.

Furthermore, it needs to be considered that electrochemical attractive forces among amino acids' side chains are highly affected by other physicochemical environmental factors. This causes the spatiotemporal flexibility in the conformation shape of proteins and subsequently their interactive affinity with other molecules. Therefore, unlike the static structure of crystals (in non-living system), 3D structures of proteins are highly flexible and dynamic. The highly dynamic conformational shape of proteins makes these molecules highly efficient logic gates for various biological purposes (please see part 4, conditional coding).

For future research, we suggest the careful design of DNA and protein-coding sequences for generation of autonomous systems. In addition, hybrid DNA-protein

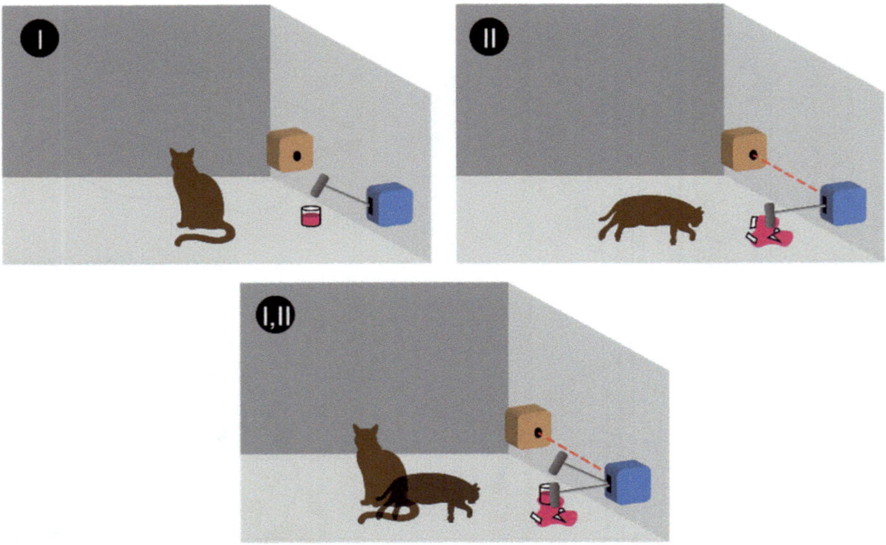

Fig. 2.13 Schematic image representative of the interpretation of quantum mechanics, initially explained by Erwin Schrodinger [26]. A cat, a flask of poison, and a radioactive source are placed in a sealed box. If an internal monitor (e.g., Geiger counter) detects radioactivity, e.g., a single atom decaying, the flask is shuttered, releasing the poison, which kills the cat. Therefore, despite the existence of the possibility of both live and dead, only one condition would be observed after opening of the box

algorithms can be applied for generation of algorithmic self-assembly by the guided design of electrochemical attractive forces among coding molecules. A detailed description of algorithmic self-assembly and autonomy is discussed in Chap. 3.

2.2.4 Inherent and Conserved Coding

A unique property of biological coding is related to the capacity of data inheritance, which means information can be transferred from one generation to the next one. DNA provides a highly stable matrix for data inheritance among different generations in a biological system. For example, DNA residuals of ancient oil-producing bacteria can be detected in oil samples (Fig. 2.14). In addition, DNA provides a highly stable media for the conservation of biological data for maintenance of species over millions of years.

2.2.5 Conditional Coding

A special property of coded molecules (e.g., DNA or proteins) is related to their capacity for direct connection and reactivity to their environmental factors. In fact, even though biological data are stored in a highly conserved and stable coded material like DNA, the coding system is still highly dynamic and able to adapt to the

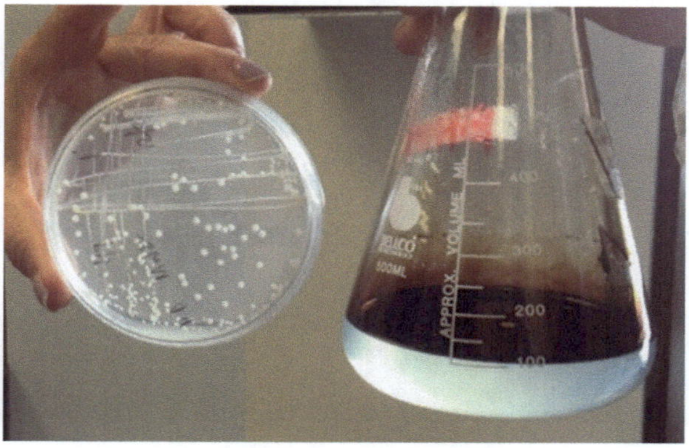

Fig. 2.14 An oil sample (right) is grown in an enriched substrate to isolate microbes; pure culture (left) of a strain of biosurfactant-producing microbes isolated from the oil sample

Fig. 2.15 Conditional coding in a plant's seeds. Initiation of growth in plants' seeds is conditional to the existence of special levels of light temperature and humidity

environmental conditions. In fact, DNA molecule is highly connected to the environmental factors and its expression always is affected by environmental conditions.

This property of biological coding for adaptation with environmental conditions here is called conditional coding. Figure 2.15 provides an example for the existence of conditional coding in a growing seed.

A DNA molecule in a seed carries the information for all biological operations and generates a complete plant, but the growth operation process will not start until the initiation of special environmental conditions including light, temperature, and specific level of humidity. This mechanism is highly regulated through intermediate molecular switches. Figure 2.16 provides schematic illustration representative of conditional coding through logic gate switches for activation of growth-signaling pathways.

Expression of amalyse (an enzyme) is critical for initiation of growth-related pathways in a seed such as metabolic and energy-producing pathways. Enzymatic activity of amylase on starch (which is the main energy storage resource for the seed) causes the release of glucose. Consequently, glucose molecules are accessible for metabolic pathways and provide the energy requirements for early stages of growth before the generation of photosynthesis machinery. Expression of the amylase gene is conditioned to the existence of light, temperature, and humidity.

Conditional coding in cells is a great model for a generation of synthetic systems with the capacity of quick adaptation with the environmental factors. Unlike current

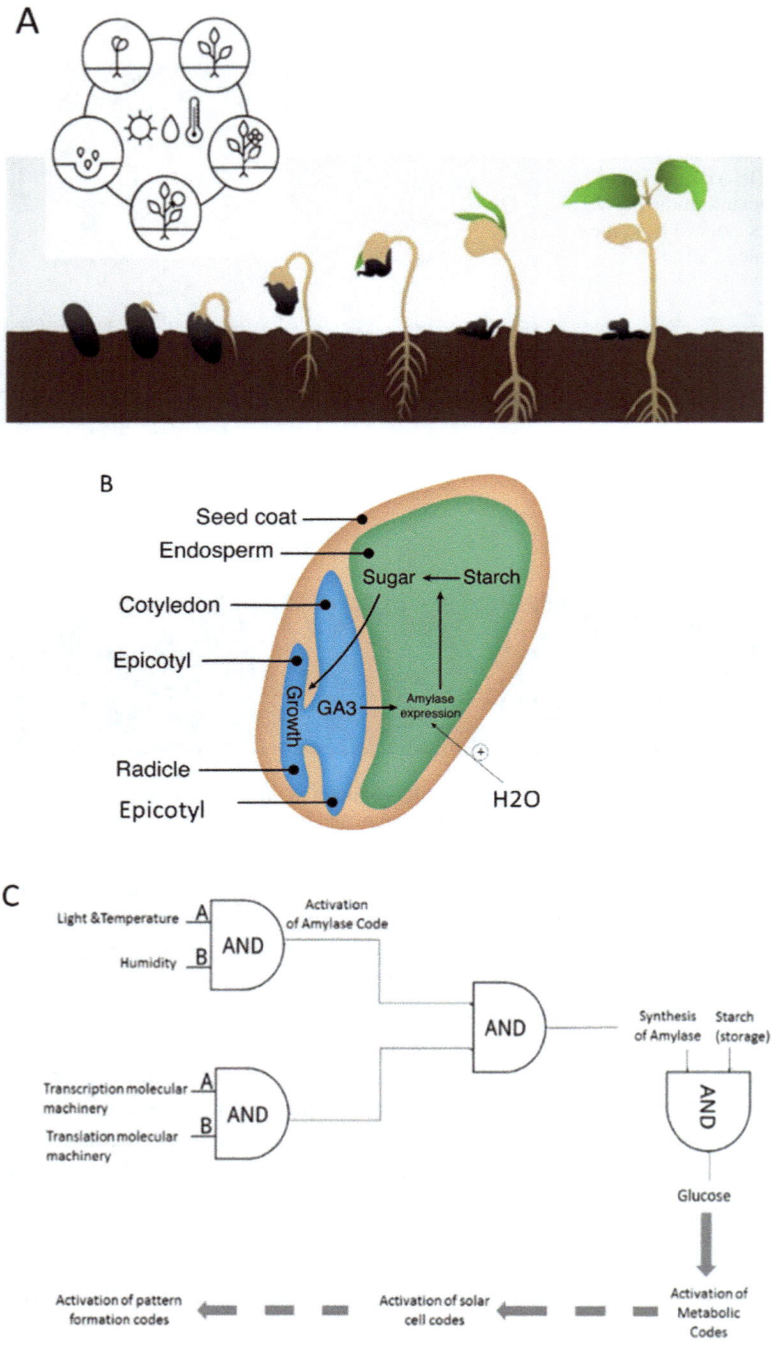

Fig. 2.16 (**a–c**) Schematic image representative of conditional coding in a plant's seed. (**a**) DNA molecule in a seed carries the information for all biological operations and generates a complete plant, but the growth operation process will not start until the existence of special environmental conditions including light, temperature, and a specific level of humidity. (**b**) Expression of amylase is critical for initiation of growth-related pathways in a seed. In fact, expression of amylase is essential for initiation of metabolic and energy-producing pathways. (**c**) Schematic representative of the function of logic gate switches for activation of growth-signaling pathways

computers, DNA involves special sequences that are not carrying any information for the synthesis of biomolecules. These sequences are called noncoding sequences. DNA noncoding sequences are specialized for direct connection with environmental factors. These noncoding sequences also are called gene regulatory sequences. Gene regulatory sequences are able to recognize and provide an appropriate response to the environmental factors through a set of intermediate molecules which function as logic gate switches.

The high level of structural flexibility of proteins and their interaction capability with each other and other biochemical components makes them highly efficient switching elements for data processing networks in biological systems. For example, in a protein-protein signal transduction and data processing network, proteins can function as switching elements or signal transducers [15, 17].

Switching capability of proteins is due to a conformational change induced by an input signal [17]. Signal transduction occurs by a switching protein (e.g., due to the phosphorylation or interactions with cAMP). In addition, the intermolecular allosteric interaction between a regulatory domain (receiving the input signal) and a functional domain (transmitting the output signal) makes up the logic gate activities of proteins. Signal-transducing proteins are components of logical gates in biological data processing systems. Switches can be used to carry out logical operations of the type NOT, AND, OR, and NOR according to the rules of Boolean algebra. These operations are sufficient to process any kind of logical information [14].

2.2.6 Dynamic and De Novo Coding

One special property of biological systems is related to their de novo coding based on their dynamic environmental condition. Benefits of understanding these mechanisms are not limited to the biomedical science as they can also be applied in machine learning and engineering of internally intelligent systems. For instance, a remarkable property of the immune system is the capacity of real-time de novo coding for synthesis of antibodies just after the antigen presentation to the body. Applying this mechanism, the immune system (even with the existence of a limited number of immunoglobulin G genes) is able to respond to the antigens that never have had a single specific gene in the genome. However, mechanisms of dynamic coding, such as alternative splicing and rearrangement of coding subunits of immunoglobulins, cause the production of highly specific antibodies against antigens that are presented to the body for the first time (Fig. 2.17).

2.2.7 Integration of Software and Hardware

Another property of biological coding is related to the unity of data storage and operating system. The entire coding and operating units in biological systems are derived from multifunctional materials that perform different functions at different stages and based on environmental conditions (Figs. 2.18 and 2.19). Unity of software and hardware will be discussed in more detail in Chap. 3.

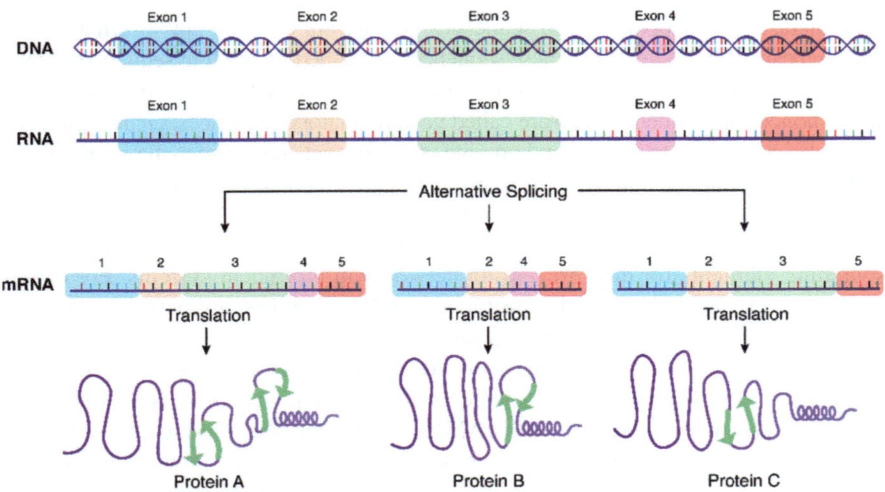

Fig. 2.17 A schematic image representative of the mechanism of mRNA alternative splicing. Alternative splicing of mRNA molecules causes the production of different types of proteins from one gene, based on the dynamic microenvironmental signals

2.3 Cognitive Chemistry Algorithm: A Biomimetic Multilayer Molecular Coding Algorithm for Solving NP Problems

In computational complexity theory Nondeterministic Polynomial time (NP) problems are a class of combinatorial optimization problems that are solvable in polynomial time by a theoretical non-deterministic Turing machine. That means, regarding the combinatorial nature of NP problems, the required time for solving the problem is a polynomial function of the size of input data. Therefore, by increase in the size of input data, the operation time for solving of problem increases exponentially. NP problems are defined as one of the most important open questions in mathematics. NP problems also are categorized as a class of decision-making problems in computational complexity theory. The main importance of NP problems is related to their vast application in designing of decision-making algorithms in artificial intelligence (AI) [9].

There have been significant efforts in conceiving computational approaches for solving NP problems applying current silicon-based and quantum computers. Living cells apply highly efficient computational methods to solve similar NP problems for different biological benefits. For example, during the early stages of embryonic development, thousands of NP problems can be operated per second in parallel to each other through the developmental signaling pathways in differentiating stem cells (Fig. 2.20). Stem cells can make thousands of decisions per second by finding the optimal combination of gene regulatory factors toward different destinations (e.g., differentiation toward various tissues and organs).

As an example of NP problems, we refer to traveling salesman problem (TSP). TSP is a combinatorial optimization problem, important in both operations research as well

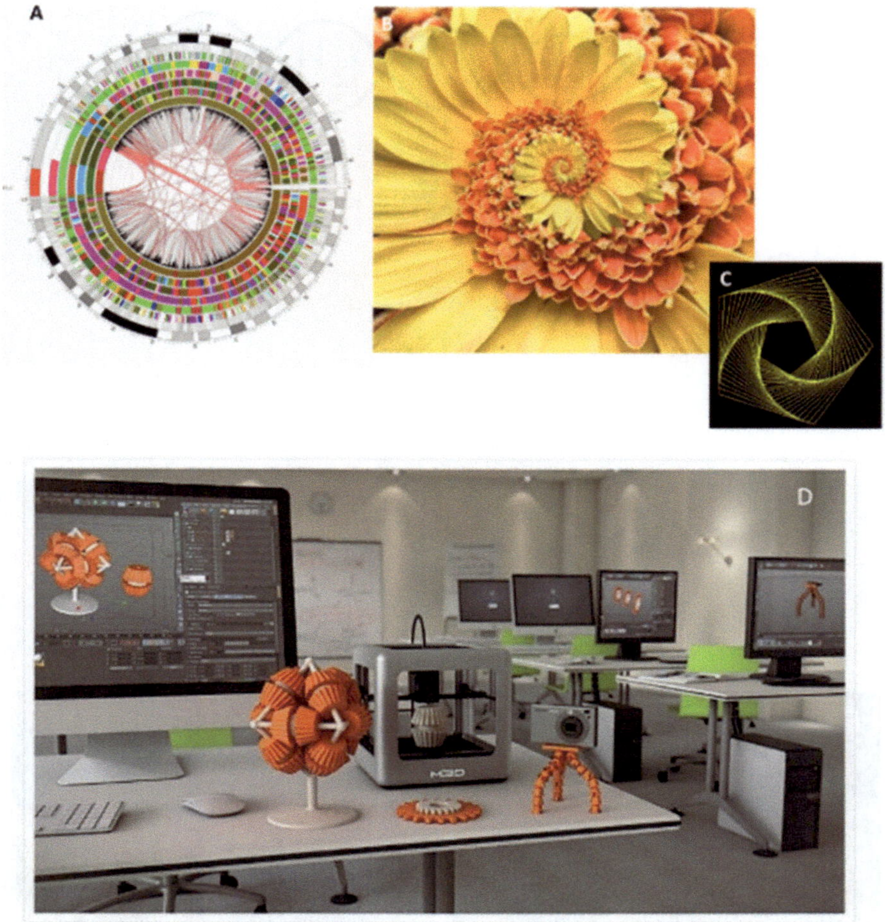

Fig. 2.18 (**a–d**) Illustration of integration of software and hardware in a multilayer molecular-based coding system. (**a**) A schematic image representative of gene regulatory networks encoding the information for generating a plant. (**b, c**) A combination of multilayer coding and molecular coding causes the unity of software and hardware, unlike the electron-based binary coding where a software needs a hardware (e.g., 3D printer) to generate a physical object (**d**)

as theoretical computer science. Briefly, the TSP asks the following question: Given a list of cities and the distances between each pair of cities, what is the shortest possible route that visits each city once and only once and returns to the origin city? [2, 8].

Figure 2.21 illustrates a TSP for an airway optimization problem. Figure 2.22 illustrates a schematic representation of a TSP with 6 cities and 15 roads. Cities are demonstrated by nodes, and roads between cities are shown as edges. Numerical values are representative of distances between each pair of cities. As it can be observed in Figs. 2.21 and 2.22, due to the combinatorial nature of the TSP problem, as the number of cities increases, the complexity of the problem increases exponentially.

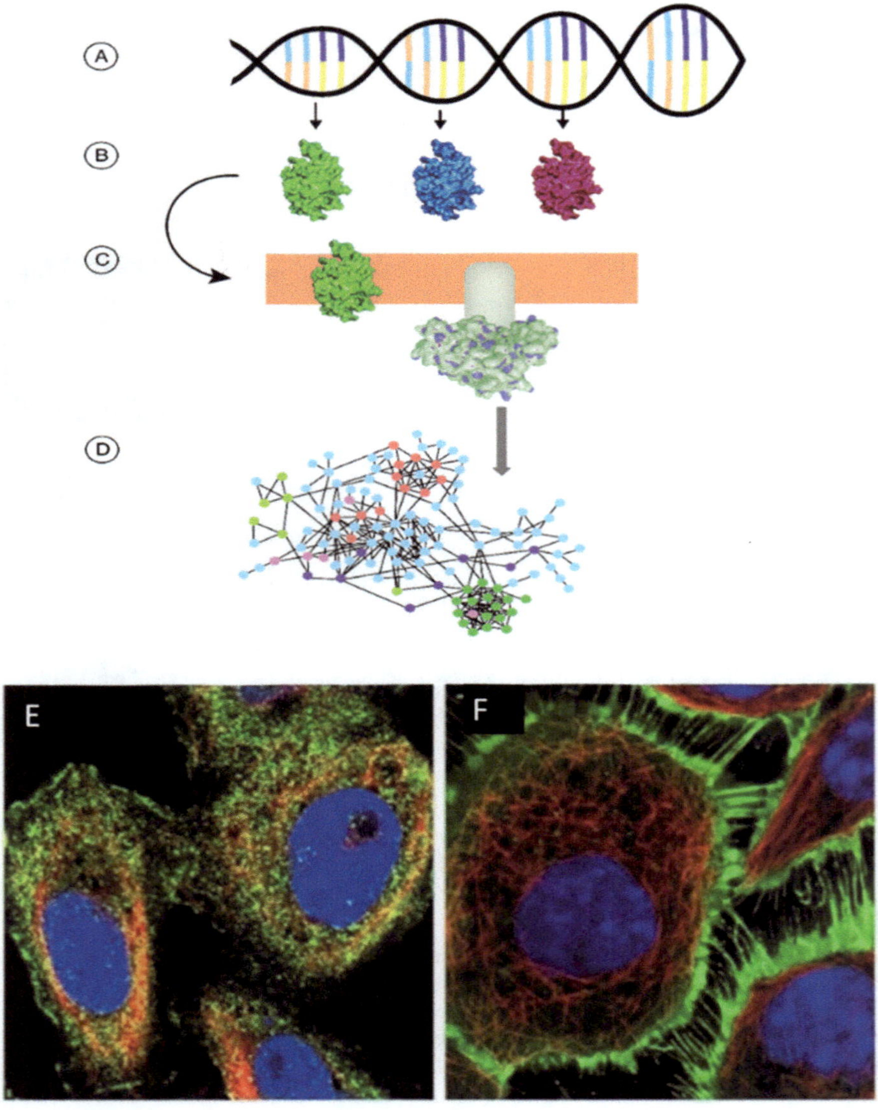

Fig. 2.19 (**a–d**) Schematic image representative of integration of software and hardware in biological systems. (**a**) DNA coding layer which encodes the information for the synthesis of a membrane signal transducer. (**b**, **c**) Schematic image illustrating how the output of the first layer of coding is an input for the next layer. For example, the output of the first layer of coding functions as a part of a protein signaling pathway as well. (**d**) Schematic image representative of protein-protein interaction signaling pathway. (**e**, **f**) IHC staining of fibroblast against a cytoplasmic signaling protein. (**e**) IHC staining against a cytoskeletal protein (red color) and a jap junction (green color) which are playing bifunctional roles in cell signal transduction (software function) as well as the formation of the physical structure of cell (hardware) at the same time

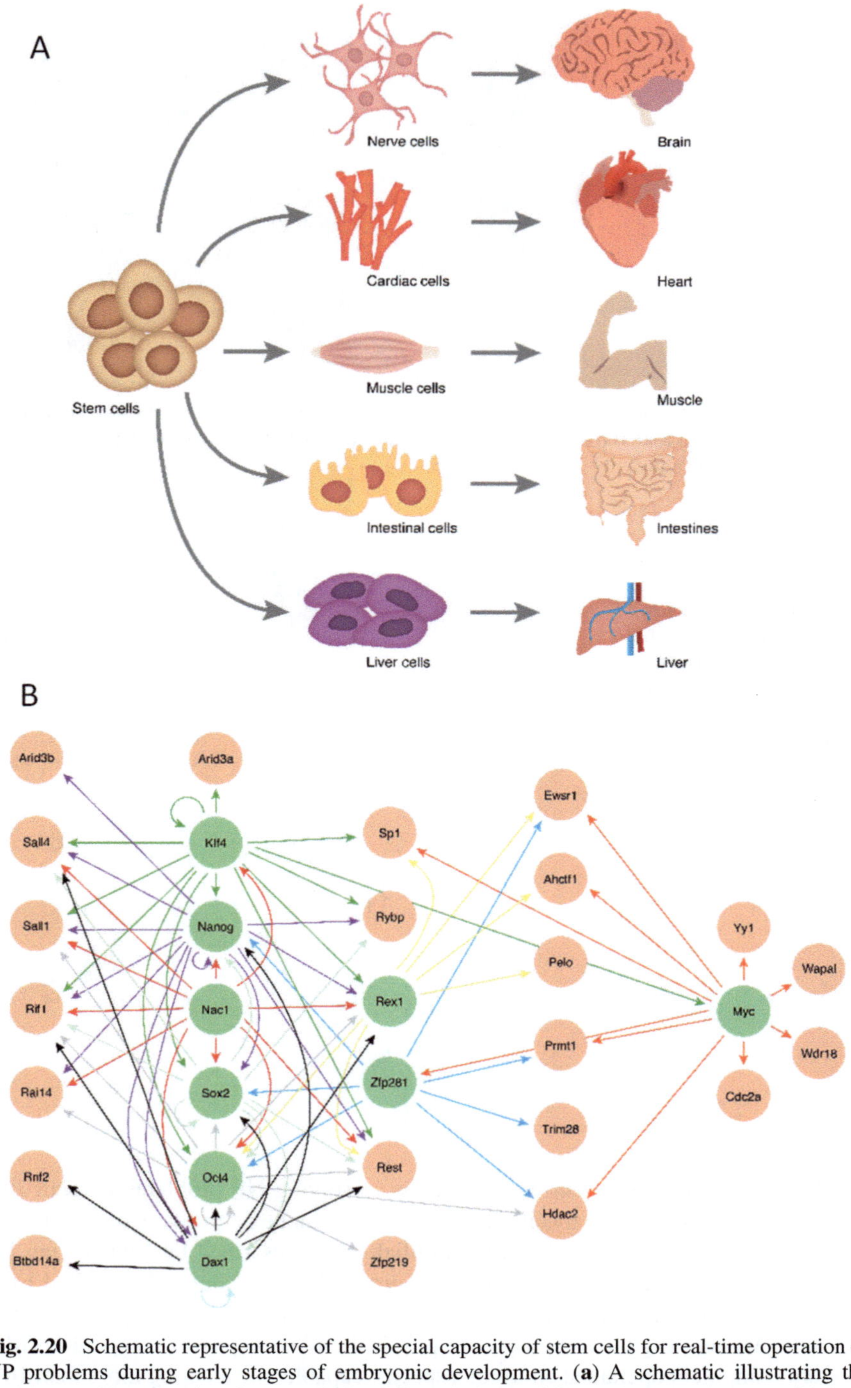

Fig. 2.20 Schematic representative of the special capacity of stem cells for real-time operation of NP problems during early stages of embryonic development. (**a**) A schematic illustrating the decision-making capacity of stem cells for differentiation to different tissues and organs. (**b**) Stem cells can make thousands of decisions per second by finding the optimal combinations of gene regulatory factors in a NP problem with a network as large as the entire of their genome

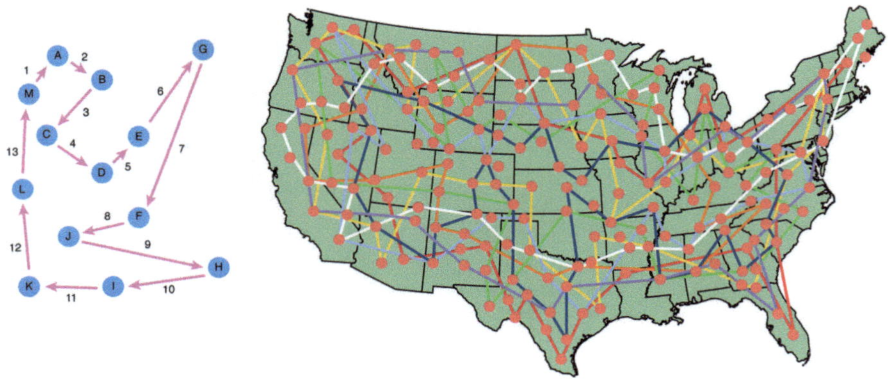

Fig. 2.21 Schematic representative of a TSP problem in airway optimization. In a given number of cities, the problem asks for the shortest possible pathway that visits each city exactly once and comes back to the original city

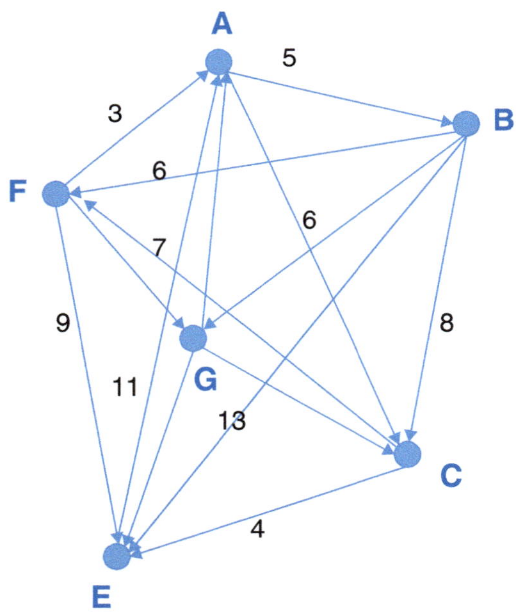

Fig. 2.22 A simple schematic representative of a TSP. Cities are shown by nodes and roads are shown by vertices between each pair of nodes. Values are representative of various distances between each pair of cities

The main limitation of current computers in solving NP problems is related to their sequential operation capacity. The theory of computer science is based on universal Turing machines (UTMs), which initially was described by Alan Turing [4, 33]. Based on the Turing theory, a task is computable if it could be specified by a sequence of instructions which results in the completion of a task applying the same machine. Modern digital computers are physical embodiments of classical UTMs. Modern computers operate at enormous speeds, capable of executing more than 10^{15} instructions per second (current fastest computer has a Linpack performance of 93

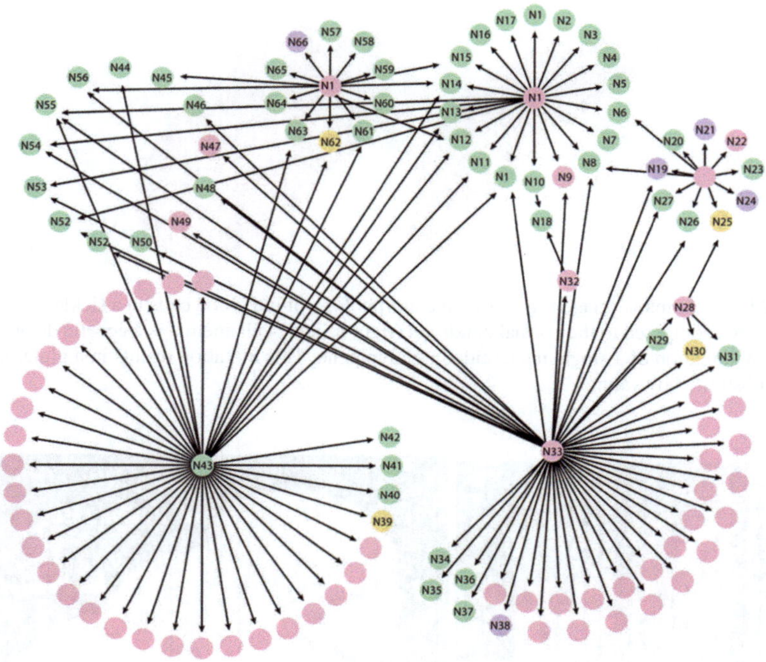

Fig. 2.23 Schematic image representative of the visualization of a local search method for solving a NP problem

peta FLOPS). However, their sequential approach to data processing is the main limitation for solving combinatorial problems such as TSP.

Current approaches for solving NP problems are mainly focused on finding the near to optimal solutions and applying local search and semi-sequential operation algorithms (Fig. 2.23) [37]. Therefore, TSP is still classified in the category of NP problems.

The high accuracy of the computational method (e.g., in finding the exact optimal solution in NP problems) is critical for biological purposes. In fact, analysis of gene expression patterns in genetic diseases compared to the normal samples indicates that near to optimal solution is not sufficient for biological purposes. In some cases, variation in one amino acid codon, over the entirety of the human genome, can cause a genetic disorder. For example, sickle cell anemia is a genetic disorder, well known by a single nucleotide mutation in B-globin gene, which results in a glutamic acid being substituted by valine. Figure 2.24 illustrates an schematic image related to changes in the morphology of red blood cell in a sickle cell anemia patient.

Furthermore, results of genetic analysis of human and chimpanzee indicate 98% homology between the human's and chimpanzee's genome. It can be concluded that even 2% difference in the entire developmental stages of a human and chimpanzee embryo can make great structural and functional variations in biological systems (Fig. 2.25).

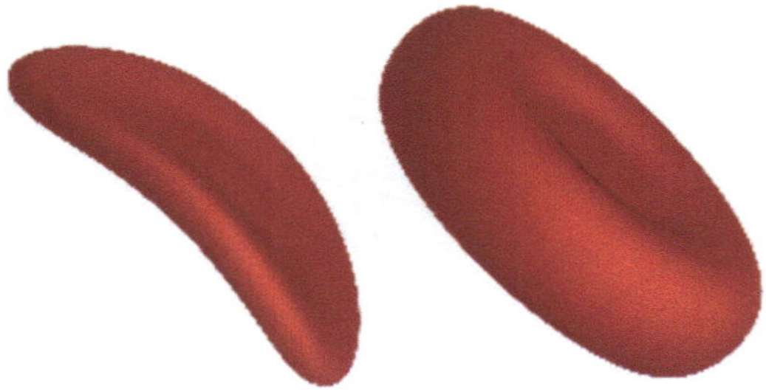

Fig. 2.24 A schematic image illustrating the morphology of red blood cells in a sickle cell anemia patient (left) compared to the normal condition (right). Sickle cell anemia is a genetic disorder that occurs by mutation of a single nucleotide in globin gene. This mutation results in a glutamic acid being substituted by valine

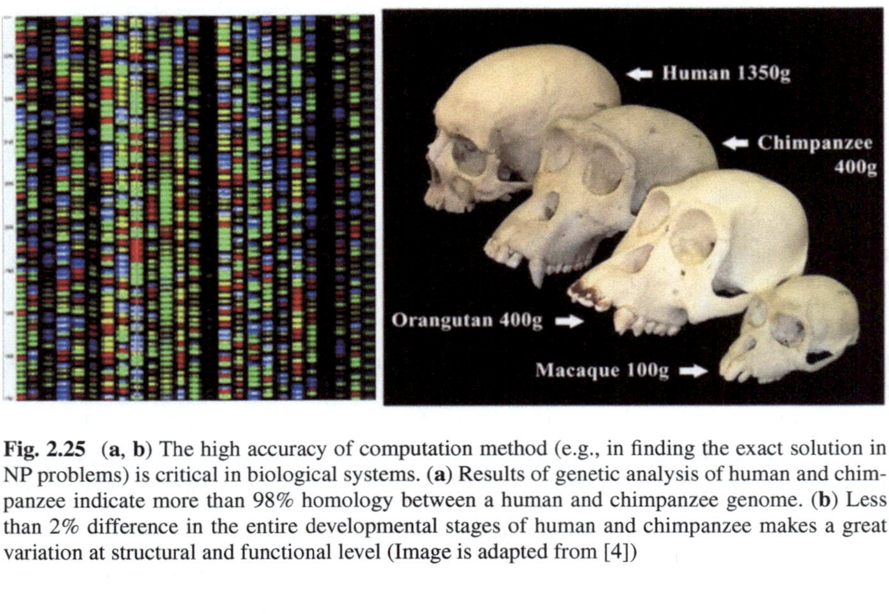

Fig. 2.25 (**a, b**) The high accuracy of computation method (e.g., in finding the exact solution in NP problems) is critical in biological systems. (**a**) Results of genetic analysis of human and chimpanzee indicate more than 98% homology between a human and chimpanzee genome. (**b**) Less than 2% difference in the entire developmental stages of human and chimpanzee makes a great variation at structural and functional level (Image is adapted from [4])

Comparative analysis of genetics in different species reveals that near to optimum solution in solving NP problems is not sufficient for biological purposes. Biological systems apply special data operation algorithms to find the exact solution for NP problems in real time.

There is still no model that provides a clear explanation for this high level of accuracy and efficiency in computational methods of biological systems.

On the other hand, recent studies on complexity problems across different industries indicated that real-life NP problems are geometrical problems with multidimensional complexity in their networks [4, 23].

In fact, current computing methods which are working based on linear operation are not sufficient for solving NP problems with geometrical complexity in their

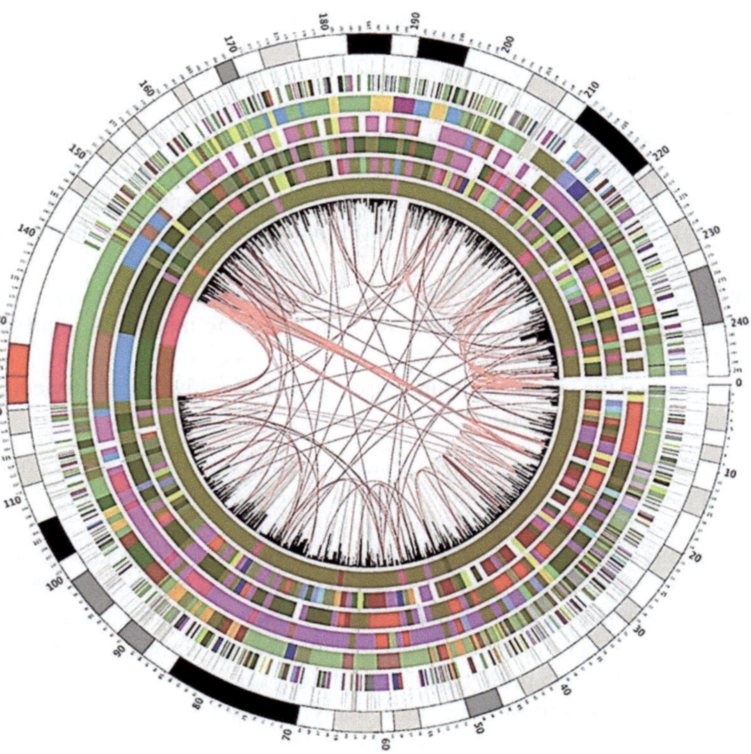

Fig. 2.26 A schematic image representative of visualization of data analysis in a biological gene regulatory network applying *CIRCOS genome visualization tools* (Image courtesy of [16]). The schematic also indicates the combinatorial nature of genomic data processing in biological systems. Genes which are representative of nodes in a TSP problem are determined by different colored barcodes. Combinatorial effect of genes on each other is determined by connecting edges among genes.

networks. Solving NP problems requires innovative approaches in designing algorithms with the nonlinear and parallel operation capacities.

Biological coding systems are promising models for designing such algorithms that satisfy the requirements of multidimensional and parallel operation to overcome the complexity of NP problems. Figure 2.26 shows a schematic image representative of a TSP problem in a complex gene regulatory network. Spatiotemporal patterns of gene expression cause the geometrical complexity of NP problems in biological systems.

Biologically inspired algorithms such as neural network [31] and ant colony [21, 38, 39] have been applied extensively for various computational purposes. At the molecular level, DNA computing has been studied initially by Leonard Adleman [1]. Since then, DNA has been the major focus of several research studies in designing biomimetic computing algorithms [1, 5, 13, 14, 18, 24, 27, 32, 39].

Other biomolecules including RNA and proteins have also been applied for computing purpose [8, 11, 18, 20, 34]. Faulhammer and coworkers reported an

RNA-based computation solution to chess problem [7]. Unger and Moult designed a molecular-based computing system in which proteins were used as NADN logic gates [34]. Nicolau et al., (2016) reported a parallel computational method with molecular- motor-propelled agents in nanofabricated protein-based networks [18].

However, there is still very limited information about the biologically inspired algorithms that could simulate multilayer coding and operation capacities of living cells. For example, gene regulatory networks or protein-protein signal transduction networks have yet to be applied as models for the generation of biomimetic computing algorithms.

Simulation of mechanisms of multilayer coding and data processing in cell signaling pathways can provide powerful tools in designing innovative algorithms for solving NP problems.

To summarize, the significance and capacities of cognitive chemistry coding and data operation system include:

- The massive information storage capacity through multiple levels of coding.
- The capacity of massive parallel operation applying numerous number of molecular operators (e.g. DNA, RNA, and Proteins).
- The capacity of autonomous data operation through the electrochemical interactions of molecules and atoms on each other which can be described through the quantum mechanic's effect of atoms on each other (e.g., the autonomous formation of three-dimensional conformation of proteins through the amino acid-amino acid interaction during protein folding).
- The capacity of multilayer operation applying different types of coding molecules and materials at different levels of complexity. Figure 2.27 illustrates schematic images representative of massive operation capacity of cognitive chemistry

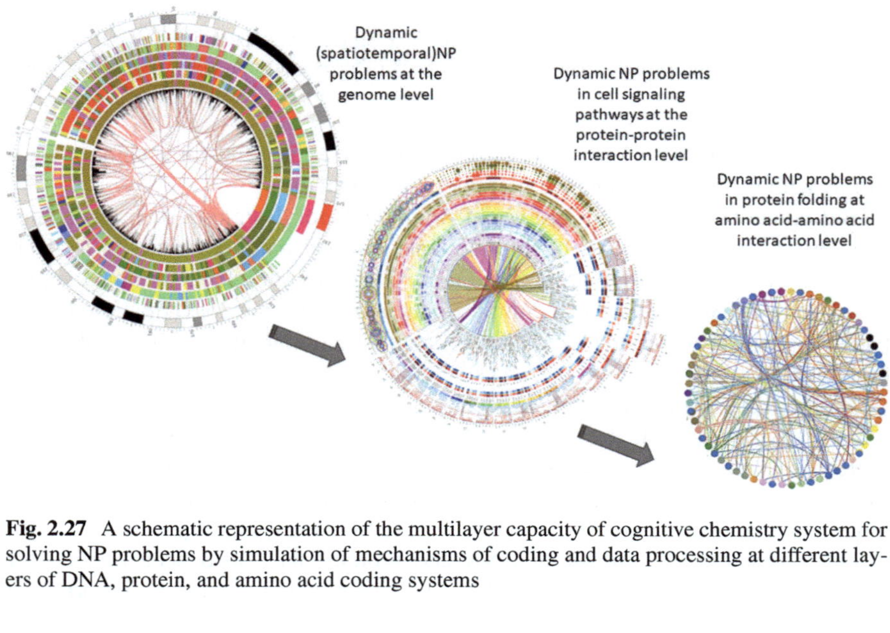

Fig. 2.27 A schematic representation of the multilayer capacity of cognitive chemistry system for solving NP problems by simulation of mechanisms of coding and data processing at different layers of DNA, protein, and amino acid coding systems

coding system for solving NP problems at different layers of complexity including gene regulatory networks, protein-protein interaction networks, as well as amino acids interaction networks during protein folding.

The autonomous and combinatorial nature of molecular interactions, at each coding and operation layer, leads to the real-time and massive operation capacity of the system in solving NP problems. Though not discussed here, the exquisite accuracy and efficiency of data operation in living cells through the highly complex and interconnected DNA-protein NP networks motivated us to develop a biomimietic computing algorithm. This algorithm is leveraged by the wide computation potentials of both nucleic acids and proteins for solving non-biological NP problems.

References

1. Adleman L (1994) Molecular computation of solutions to combinatorial problems. Science 266(5187):1021–1024
2. Afaq H, Saini S (2011) On the solutions to the traveling salesman problem using nature-inspired computing techniques. IJCSI 8:326–334
3. Allard A, Serrano MA, Garcia-Perez G, Boguna M (2017) The geometric nature of weights in real complex networks. Nat Commun 216:1–8
4. Bradbury J (2005) Molecular insight in to human brain evolution. PLoS Biol 3:0367–0370
5. Chen YJ, Dalchau N, Srinivas N, Phillips A, Cardelli L, Soloveichik D, Seelig G (2013) Programmable chemical controllers made from DNA. Nat Nanotechnol 8:755–762
6. De Castro LN (2007) Fundamentals of natural computing: an overview. Phys Life Rev 4:1–36
7. Faulhemmer D, et al. (2000) Molecular computation: RNA solution to chess problems. PNAS, USA 57(4):1385-1389
8. Garey MR, Johnson DS (1979) Computers and intractability: a guide to the theory of NP completeness. W. H. Freeman & Company, New York, U.S.A
9. Feng C et al (2013) Codon usage patterns in Chinese bayberry (Myrica rubra) based on RNA sequencing data. BMC Genomics 14:732
10. Hug H, Schuler R (2001) Strategies for the development of a peptide computer. Bioinformatics 17:364–368
11. https://simple.wikipedia.org/wiki/Travelling_salesman_problem
12. Kim J et al (2008) An extended transcriptional net-work for pluripotency of embryonic stem cells. Cell 132:1049–1061
13. Lee JY, Shin SY, Park TH, Zhang BT (2004) Solving traveling salesman problems with DNA molecules encoding numerical values. Biosystems 78:39–47
14. Liu Q et al (2000) DNA computing on surface. Nature 403:175–179
15. Mark F, Klingmuller U, Decker K (2009) Cellular signal processing, an information to the molecular mechanism of signal transduction. USA, Gaelan Science, Tylor and Francis Group
16. Naquin D, Aubenton-Carafa Y, Thernes C, Silvain M (2014) Circus: a package for circus display of structural genome variation for paired-end and mate-pair sequencing data. BMC Bioinformatics 14:198
17. Nelson DL, Cox M (2017) Lehninger principles of biochemistry, 7th edn. W.H. Freeman & Company, New York
18. Nicolau D Jr et al (2016) Parallel computation with molecular- motor- propelled agents in nanofabricated networks. PNAS 13:2591–2596
19. Qian L, Winfree E (2011) Scaling up digital circuit computation with DNA strand displacement cascades. Science 332:1196–2011

20. Redriguez RA, Yu L, Chen LY (2015) Computing protein- protein association affinity with hybrid steered molecular dynamics. J Chem Theory Comput 11:4427–4438
21. Roy S (2013) Bioinspired ant algorithms, a review. J Modern Education Comput Sci 4:25–35
22. Rubens J, Selvaggio G, Lu TK (2016) Synthetic mixed signal computation in living cells. Nat Commun 2016(7):11658
23. Rune J et al (2015) Identifying causal gateways and mediators in complex spatiotemporal systems. Nat Commun 6:8502
24. Schatz MC, Langmead B, Sazberg S (2010) Cloud computing and DNA data race. Nat Biotechnol 28:691–693
25. Schrodinger E (1944) What is life. Cambridge University Press, UK
26. Schrodinger E (1935) Die gegenwartige Situation in der Quantenmechanik. Die Naturwissenschaften 23 (48):807–812
27. Shapiro E, Ran T (2013) DNA computing: molecules reach consensus. Nat Nanotechnol 8:703–705
28. Singh S, Lodhi EA (2013) Study of variation in TSP using genetic algorithm and its operator comparison. IJSCE 3:2231–2307
29. Siuti P, Yazbek J, Lu TK (2013) Synthetic circuits integrating logic and memory in living cells. Nat Biotechnol 2013(31):448–452
30. Tarkov MS (2015) Solving the traveling salesman problem using a recurrent neural network. Am Anal Appl 8:275–283
31. Tulpan D (2014) Thermodynamic post processing versus GC- content pre- processing for DNA codes satisfying hamming distance and reverse-complement constraints. JEEEA ACM Trans Comput Biol Bioinform 11(2):441–452
32. Turing A (1936) On computable numbers with an application to Entcheidung problem. Proc Lond Math Soc II Ser 42:230–265
33. Unger R, Moult J (2006) Towards computing with proteins. Proteins 63(53–64):9
34. Wang Z, Dongmei H, Meng H, Tang C (2013) A new fast algorithm for solving minimum spanning tree problem based on DNA molecules computation. Biosystems 1114:1–7
35. Weissman JA, Pan YA (2013) New resource and emerging biological application for multi-color genetic labeling analysis. Genetics 199(2):293–306
36. Weissman JA et al (2011) Generating and imaging multicolor Brainbow mice, Cold Spring Harbor Laboratory Protoc. https://doi.org/10.1101/pdb.top114
37. Wong L, Low MYH, Chong CS (2010) Bee colony optimization with local search for traveling salesman problem. Int J Artif Intell Tools 19(3):305–334
38. Yang J, Dung R, Zhang Y, Cong M, Wang F, Tang G (2015) An improved ant colony optimization (I-ACO) method for the quasi traveling salesman problem (Quasi-TSP). Int J Geogr Inf Sci 29:1534–1551
39. Zhang M, Cheng M, Tarn JA (2006) Mathematical formulation of DNA computation. IEEE Trans Nanobioscience 5(1):32–40

Chapter 3
Molecular Mechanism of Self-Organization in Biological Systems

Abstract Self-organization is the autonomous formation of complex structures from units of less complexity by local internal interactions. As a precise model of self-organization in nature, embryonic organogenesis is a process where different tissues and organs form in a growing embryo by the autonomous assembling of cells.

Embryonic self-organization, which occurs at different levels of complexity from nano to macro levels in biological systems, is a highly efficient autonomous process. Self-organization is like robotics without wires and motors. It means the manufacturing program is embedded in materials themselves.

When we look at the manufacturing and construction process through different industries, there are major efficiency issues in energy consumption and labor work compared to autonomous formation and assembling of system's parts during self-organization.

In this chapter, we will discuss the regulatory mechanisms behind embryonic organogenesis through information storage in biomolecules. In addition, specifically, we discuss the molecular regulation of both differentiation and morphogenesis and their application in regenerative medicine. Deep understanding of regulatory mechanisms of self-organization can open new avenues in designing next generation smart, self-organizing materials and systems for both industrial and biomedical applications.

Keywords Self-organization · Embryonic organogenesis · Morphogenesis · Protein folding · Autonomous manufacturing · Programmable materials · 3D and 4D printing · Cognitive chemistry

T. Karimi, *Molecular Mechanisms of Autonomy in Biological Systems*,
https://doi.org/10.1007/978-3-319-91824-2_3

"It is difficult for us in the twenty-first century to realize how totally mysterious embryological development has been" [5].

3.1 An Introduction to Self-Organization and Embryonic Organogenesis

The above cartoon shows the fundamental principles of autonomy in embryonic self-organization (an internally programmed process), compared to the generation of an organized construct by a 3D printer (an externally programmed process).

During embryonic self-organization, a complex and exquisitely ordered body is generated from the relatively unorganized fertilized egg. "It is difficult for us in the twenty-first century to realize and explain all molecular aspects of self-organization process that happens in growing embryo at early stages of development" [5]. Much remains to be elucidated about the relationships among the complex molecular systems of living entities, how they are constrained by the system as a whole as well as by physical laws. "Indeed, it is still an open question for some as to whether we have yet a sufficiently rich understanding of the laws of nature or whether we need to seek deep laws that lead to order and organization" [8–10]. "The fourth law of thermodynamics about such phenomena may ultimately be needed" [10].

Self-organization is the autonomous formation of complex structures from units of less complexity by local internal interactions of systems subunits [13, 19, 26]. Embryonic organogenesis in animals and germination in plants' seeds are precise models of self-organizing systems in nature [26, 31].

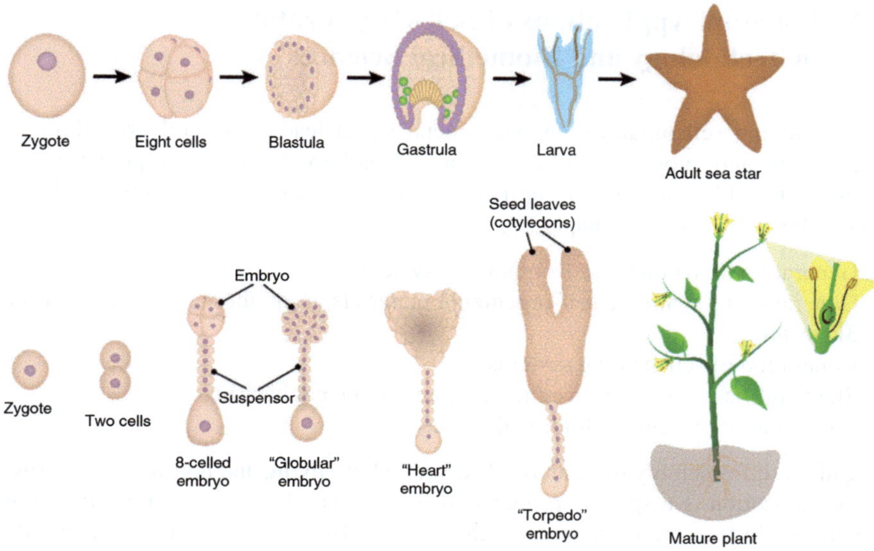

Fig. 3.1 (**a, b**) Schematic image representative of examples of self-organization in biological systems. Spontaneous formation of highly organized geometrical structures during the organogenesis in an animal fertilized egg (**a**) and germination of a plant's seed (**b**)

Figure 3.1 shows the fundamental phenomenon of autonomy during self-organization in a fertilized egg and a plant's seed. Organogenesis in animals is a process by which different types of cells, tissues, and organs gradually form at early stages of embryonic development. In plants, organogenesis occurs continuously and only stops when the plant dies. For example, formation of lateral leaves, flowers, and branches in the shooting continues constantly during the plant's lifetime [31]. Organogenesis is an autonomous process which means the instructions and programs for assembling of system's subunits are embedded in materials of the system by themselves [6, 19, 26]. In fact, organogenesis in a living form is a unique autonomous processes which integrates the units of software (instructions and programs for the progress of process) and hardware (materials that constructs the system's physical mass) all together. In other words, there is no gap between the software and hardware.

Furthermore, beyond the autonomous assembling of programmable materials, embryonic development is accompanied by computational capacity of materials.

Self-organization is like robotics without wires and motors. It means the manufacturing program is embedded in materials themselves. This is the ability to program physical materials to change shape and physical properties. In addition to the formation of structures, during the embryonic organogenesis, thousands of functional and machinery units form constantly at multiple levels of complexity (starting from molecular to micro and macro scales) at different tissues and organs to operate a collectively uniform system. This collective connection between the molecular subunits of a living organism functions as molecular algorithm which runs all developmental operations during the embryonic organogenesis.

3.2 Potential Applications of Self-Organization in Technology and Biomedical Sciences

Understanding self-organization phenomena could lead to several breakthrough applications and improvements to existing technology. Some of the potential use cases are listed below. Further applications of self-organization in biomedical science will be discussed in Chap. 6.

- Generation of synthetic self-organizing systems
- Generation of synthetic self-organizing materials as an alternative approach for 3D printing
- Generation of self-healing systems
- Stem cell-based regenerative medicine and tissue engineering
- Treatment of protein misfolding diseases

Unlike self-assembly that can be observed in both living and non-living systems; self-organization is a specific property of life and can be observed only in living organisms. In fact, there are several features in self-organizing systems which distinguish them from self-assembly, including structural complexity and energy balance of the system. In the next section, we will discuss the main properties of self-organization and how it can be distinguished from self-assembly.

3.3 Remarkable Properties of Self-Organization

3.3.1 Polarity Formation

One of the most remarkable properties of self-organization is the capacity of polarity formation, which also can be defined as the symmetry breaking capacity of the system [20].

Polarity formation involves autonomous generation and assembling of highly specific geometrical structures, which can grow asymmetrically through different layers of complexity from nanostructures to micro- and macrostructures (Figs. 3.1 and 3.2) [13]. Therefore, self-organization is accompanied by the formation of geometrical structures with highly specific patterns and morphologies. The process also is called morphogenesis.

Figure 3.1 illustrates examples of polarity formation in growing animal and plant embryos. Figure 3.2 compares the special features of self-organization (including the polarity formation and generation of multilayer structural complexity) in a plant versus the uniform molecular organization of self-assembling crystals. Figure 3.3 illustrates the autonomous reorientation of cells and establishment of body axes in a growing amphibian embryo [2, 11]. Polarity formation in a self-organizing system requires a precise internal program and a source of energy to guide and enforce the progress of the process. Thus, self-organization only occurs through a dedicated internally programmed process [27, 31].

Fig. 3.2 (**a**, **b**) Comparison of self-organization and self-assembly. Polarity formation and multi-layer structural complexity formation in a self-organized plant (**a**) compared to the uniform molecular structure of a self-assembling crystal (bismuth crystal)

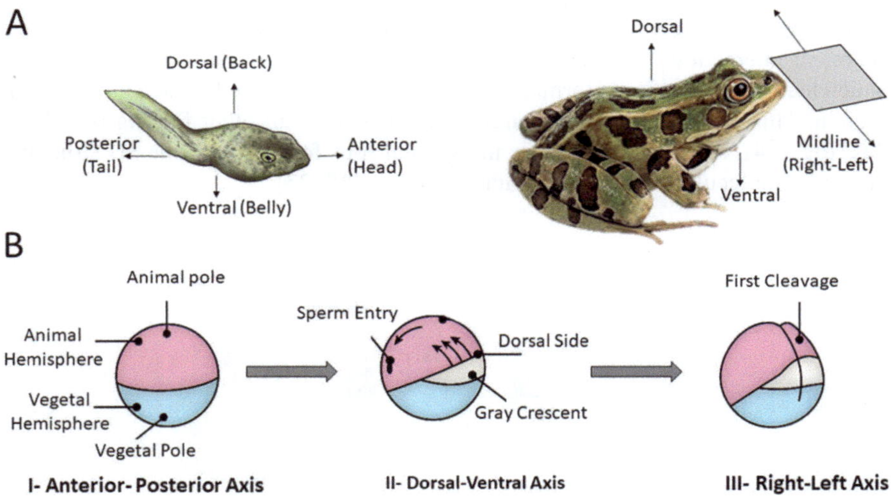

Fig. 3.3 (**a**, **b**) Gradual polarity formation and generation of body axis in a growing amphibian embryo. (**a**) Schematic image illustrates the three axes over the body of a developed embryo. (**b**) Schematic image illustrates the gradual polarity formation, reorientation of cells, and establishment of body axes during the early stages of embryonic development

3.3.2 Energy Balance

Unlike self-assembly that can be observed in both living and nonliving systems (e.g. formation of ice crystals), self-organization can be observed only in living systems. Self-assembly can be explained by thermodynamics and through the stability of the molecular structure [1]. However, the formation and maintenance of a self-organizing system rely on continuous energy input into the system. Thus, by considering the energy balance of the system, self-organization requires the autonomous energy uptake from its environment. As another remarkable natural property, biological systems are self-fueling and can autonomously manage their energy requirements [20]. Mechanisms of energy management and self-fueling properties of biological systems are discussed in Chap. 4 in detail.

3.4 Self-Organization Is a Bottom-Up Process

Self-organization is a bottom-up process which starts from the molecular level and progresses toward micro and macro levels of complexity. Protein folding is a great example of self-organization at the molecular level. Protein folding is a process in which molecules of proteins obtain their 3D structure spontaneously [16, 17, 28]. In fact, proteins do not need an external scaffold to obtain their complex structure; instruction for the 3D structure of proteins is embedded in the sequences of amino acids. Specifically, chemical codes are embedded in the carefully ordered electrochemical interactions between chemical groups in the side chains of amino acids which lead the spontaneous formation of a unique 3D structure for each molecule of protein. This is the most stable condition compared to all other folding possibilities (Fig. 3.4). Figure 3.5 illustrates the spontaneous formation of the 3D structure of insulin molecule which is a repeatable process [18, 23].

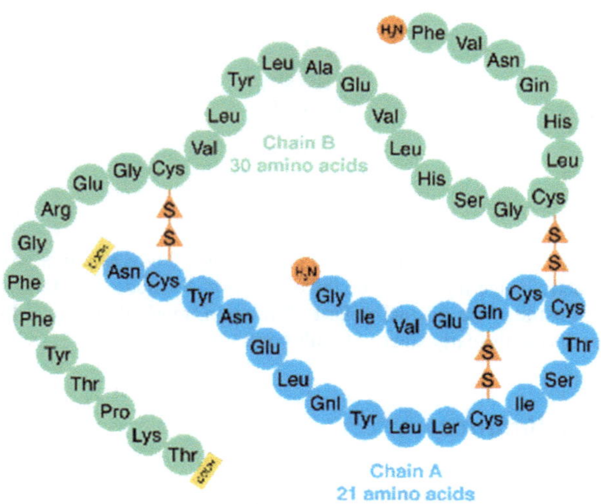

Fig. 3.4 Illustration of the spontaneous folding of insulin molecule and formation of its unique 3D structures

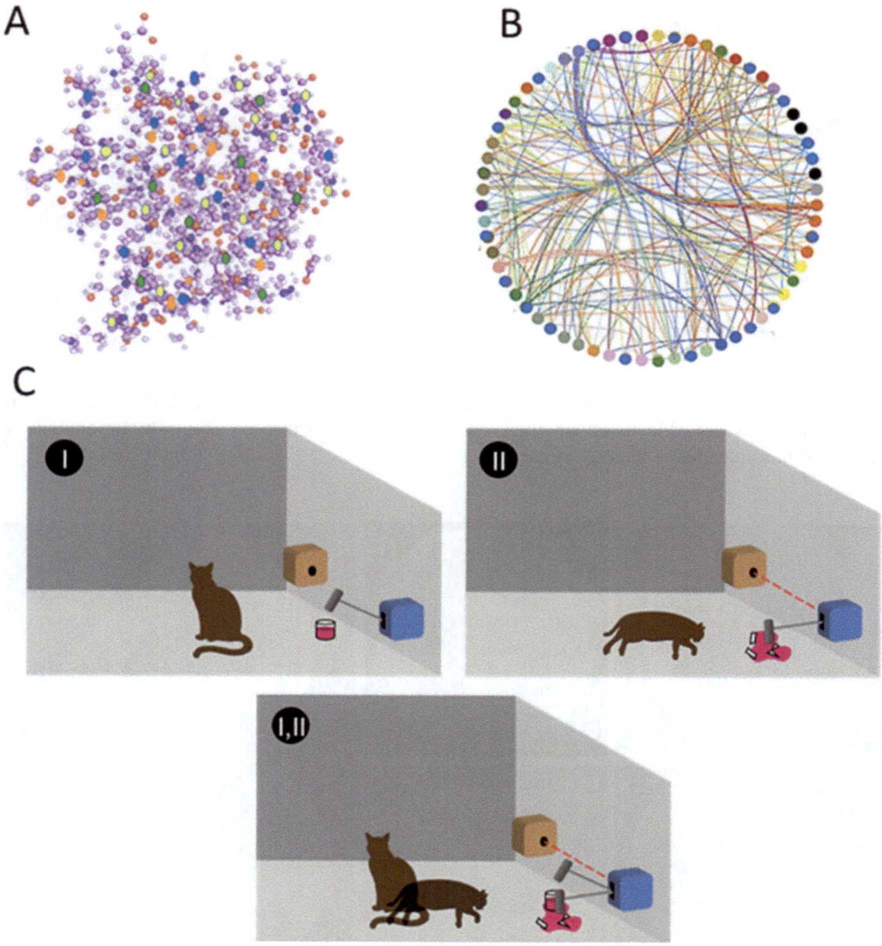

Fig. 3.5 (**a–c**) The schematic image illustrates the efficiency of autonomous and repeatable protein folding through electrochemical attractive forces among atoms (**a**, **b**). Instantaneous occurrence and repeatability of protein folding can be compared with quantum theory; meaning despite the existence of all different types of molecular interaction and large number of possible senarioes during protein folding, only one condition happens in real time which is the most stable structure

As previously mentioned, self-organization is a bottom-up process which progresses through multiple layers including nano, micro, and macro levels of complexity [11]. Figure 3.6 provides an example of self-organization at the molecular level in a protein molecule.

Figure 3.7 refers to the progress of self-organization at the levels of cells, tissues, organs, and the entire body of an organism. At the cellular level, during organogenesis, embryonic stem cells are able to assemble to make all complex structures in different tissues and organs.

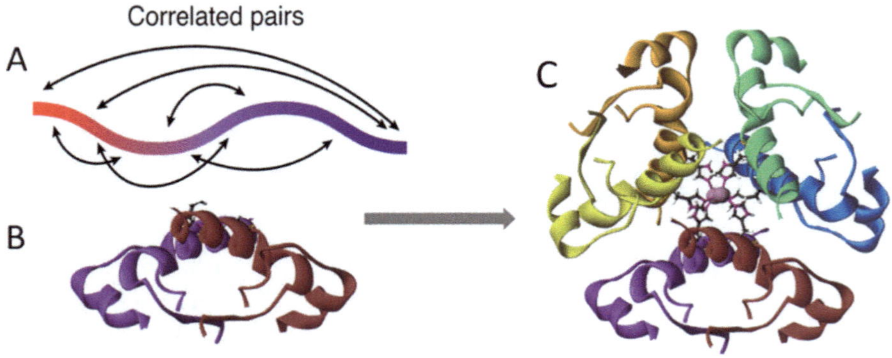

Fig. 3.6 Illustrates an schematic image representative of sequences of events (A-C) during protein folding. Proteins are able to obtain their conformational structure by the information that is instructed in sequences of their amino acids

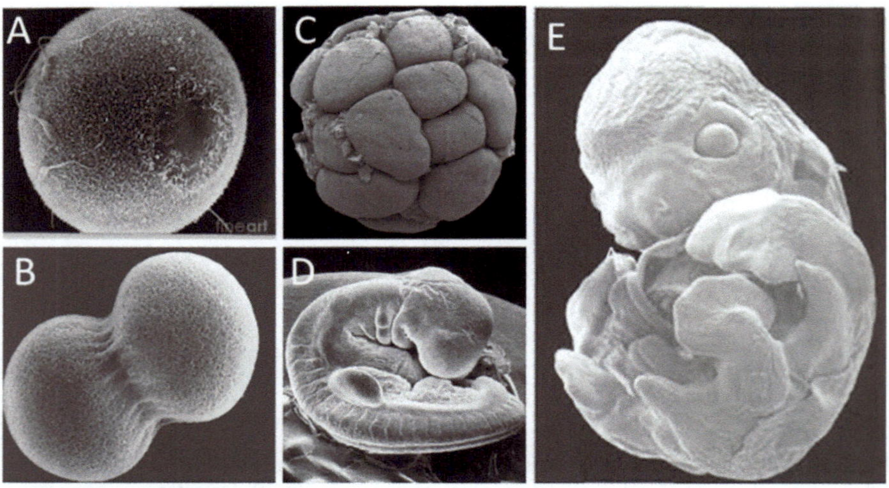

Fig. 3.7 Schematic image representative of self-organization of stem cells during the early stages of embryonic development. Stem cells are able to assemble and make highly specific structures, which grow in size and complexity by time. Scanning micrographs of the human embryo at different stages (A to E) of development (Carnegie stages). (Images are reproduced with the permission of Dr. Kathy Sulik, from collection of Dr. Vekemans and Tania Attie-Bitach. Source of images: https://embryology.med.unsw.edu.au/embryology/index.php/Carnegie_stage_table)

When we study embryonic organogenesis through multiple layers of complexity, we realize that more than structure formation, self-organization in biological systems is accompanied with more complex functions, including individual and collective cell migrations towards specific destinations, polarity formation and continues diversity formation in the primary population of cells.

Unlike self-assembly which can be explained simply by electrochemical interactions of molecules through thermodynamics, operation of more complex functions of self-organization (e.g. cell mitigation, and polarity formation) requires precise internal programs to guide the system. As we discussed in chapter 2, all cell functions are operated by molecular circuits.

Most of computational functions of cells are operated by protein signaling networks. In fact, proteins can execute nearly all operations in a cell.

Thus, enzymes provide the intricate molecular surfaces which function as microchips in a cell. Other proteins function as sensors, recognition modules and signal transducers. In addition, 4D structural flexibility of proteins enables their logic gate functionality, and provides a unique capacity to respond to the environmental factors directly. This leads to a very high computational capacity for proteins as individually operational units and in connection with other proteins and signaling molecules in dynamic cell signaling networks.

In the next part we discuss the role of cell molecular regulation behind two critical processes that occur during embryonic self-organization, including, 1- pattern formation and 2- cellular specification and differentiation.

3.5 Developmental Biology of Embryonic Organogenesis

3.5.1 Developmental Biology of Autonomous Morphogenesis and Spontaneous Pattern Formation

Morphogenesis is originated from two Greek names including morphe (means, shape) and Genesis (creation). It is a process that causes an organism to develop its shape. The process controls the organized spatial distribution of cells during the embryonic development of an organism. More than additive assembling of structural subunits, embryonic organogenesis is accompanied by more complex behaviors and functions including, 1- individual and collective mitigation of cells towards special destinations 2- commitment, specification and differentiation of initial stem cells towards different cell lineages. Morphogenetic responses may be induced in organisms by hormones, environmental chemical factors including substances produced by other cells, signaling molecules with the maternal origin, or by mechanical stresses induced by spatial patterning of the neighboring cells. Furthermore, morphogenesis is accompanied by polarity formation in the conformational shape of a growing embryo [5].

Formation of general polarity in the embryo is instructed by the embryo's genome. However, the egg cytoplasm also plays a critical role in determining the patterns of cleavage and cell specification at the early developmental stages. Once fertilization is complete, the development of a multicellular organism continues by a process called cleavage, a series of mitotic divisions whereby the enormous volume of egg cytoplasm is divided into a large number of smaller cells. In fact, fertil-

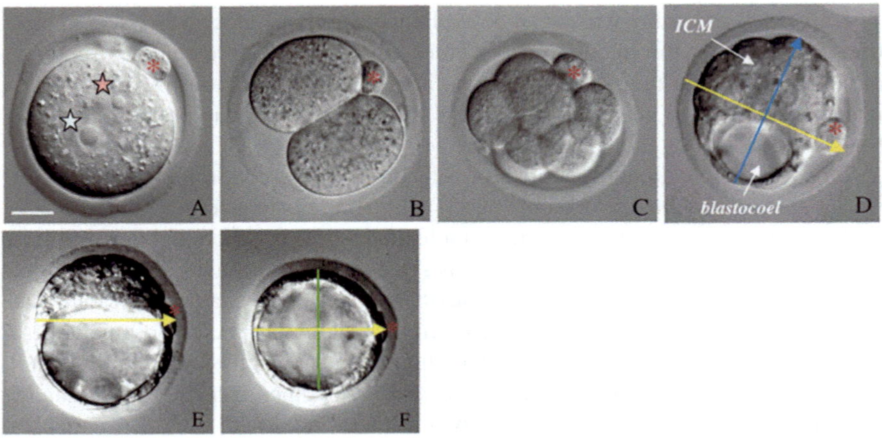

Fig. 3.8 (**a–f**) Cleavage and polarity and axis formation at early stages of embryonic development (Image adapted from Zernicka-Goetz [32])

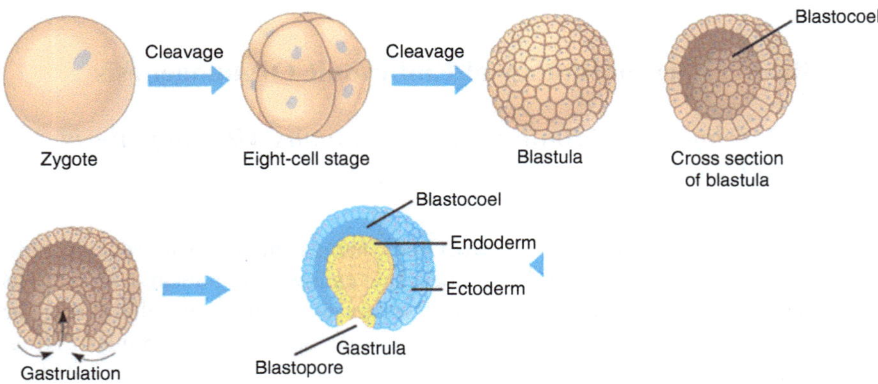

Fig. 3.9 Schematic image representative of movement of cells during gastrulation and gradual polarity formation from the nonpolar structure of blastula toward the polar structure of gastrula [5, 24]

ization gives the organism a new genome and rearrangement of its cytoplasm. These cleavage-stage cells are called blastomeres [5, 13] (Fig. 3.8).

Fast-dividing embryonic cells then later undergo spontaneous displacements and begin to move and migrate toward different parts of the embryo and acquire new neighbors. The displacement and migration of dividing embryonic cells lead the formation of early embryonic polar structures including the blastula and gastrula (Figs. 3.8 and 3.9). The blastula consists of numerous embryonic stem cells, the positions of which were established during cleavage [21].

In the next stage which is gastrulation, multilayered body plan of organism forms. Migration and rearrangement of cells during gastrulation causes the formation of three germ layers including ectoderm, mesoderm, and endoderm (from the lateral to the medial part of embryo's body) (Fig. 3.11).

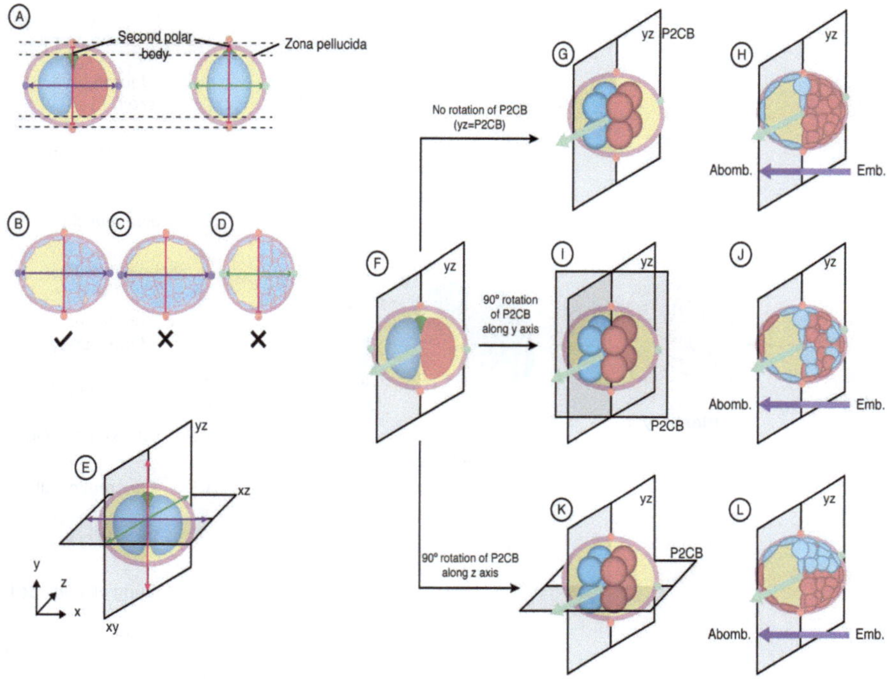

Fig. 3.10 **(a–k)** Schematic image representative of the axis by collective movement (migration) of embryonic stem cells and formation of blastula. (Image is adapted from Rossant and Tam [24])

The cells that will form endodermal and mesodermal organs are brought inside of the embryo, while the cells that will form the skin and nervous system are spread over its outside surface (Fig. 3.11). Thus, the three germ layers – outer ectoderm, inner endoderm, and interstitial mesoderm – are first produced during gastrulation. In addition, the stage is set for the interactions of these newly positioned tissues [3, 5, 21] (Fig. 3.11).

Based on results of studies on embryonic development, we know that different organisms undergo cleavage in distinctly different ways. While cleavage always precedes gastrulation, axis formation in some species can begin as early as oocyte formation (Fig. 3.10). The pattern of embryonic cleavage peculiar to a species is determined by two major parameters: (1) the amount and distribution of yolk protein within the cytoplasm and (2) factors in the egg cytoplasm that influence the angle of the mitotic spindle and the timing of its formation. During cleavage and gastrulation, the major axes of the embryo are determined and the embryonic cells begin to acquire their respective fates [5, 21] (Figs. 3.10 and 3.11).

The challenging question is: what is the regulatory mechanism behind polarity and pattern formation during and after gastrulation? Which factors are regulating the single and collective migration of embryonic cells toward the formation of specific morphologies?

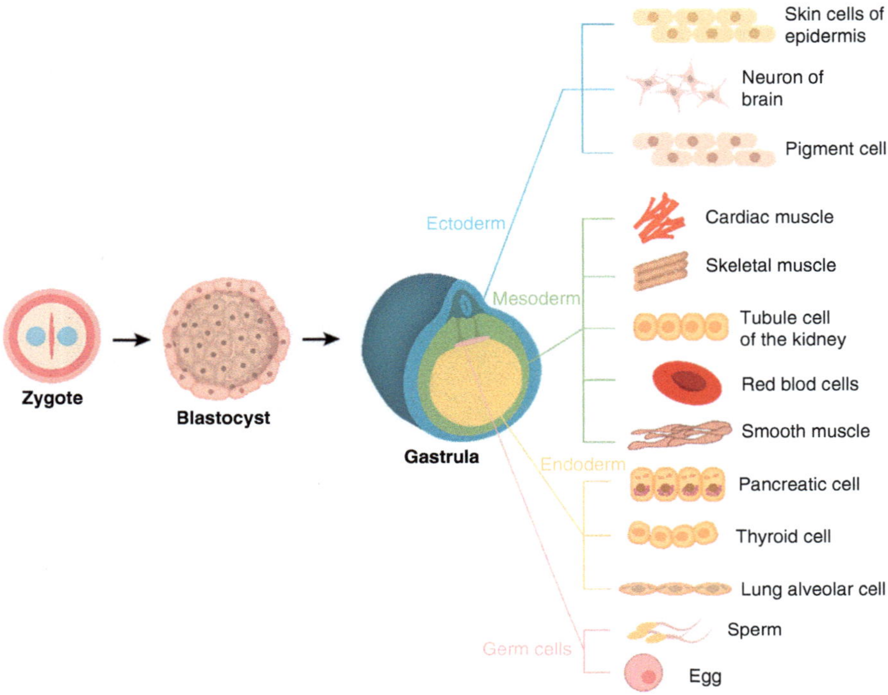

Fig. 3.11 Schematic image representative of formation of three germination layers as well as differentiation into all different cell types in the body

Cell fates are specified by an asymmetric distribution of patterning molecules. These patterning molecules also are called morphogens. A morphogen is a biomolecule that causes the fate determination of a cell by its concentration gradient.

Morphogens can be transcription factors produced within cells. They can also be paracrine factors that are produced in one group of cells and then travel to another population of cells. Various combination and concentration of morphogens can produce different biological signals for developing stem cells. Concentration gradients of morphogens produce different signals for the exposed cells, which cause the specification of the target cells [25] (Fig. 3.12). Furthermore, concentration gradient of morphogens causes the polarity formation at different parts of embryo's body.

Morphogenesis as an old mystery in the history of science, in addition to the biologists, has drawn the attention of researchers in other fields of science. Alan Turing's accomplishments in computer science are well known, but lesser known is his research interests in understanding the algorithms behind the spontaneous morphogenesis in biological systems. In his only published paper on biology, Turing proposed a mathematical model of morphogenesis. Turing offered a theory for morphogenesis based on the chemistry and through a process called intercellular reaction-diffusion [29].

In this model, a set of chemicals react with each other and diffuse across space, between cells in an embryo [5, 29, 30]. These chemical reactions need an inhibitory

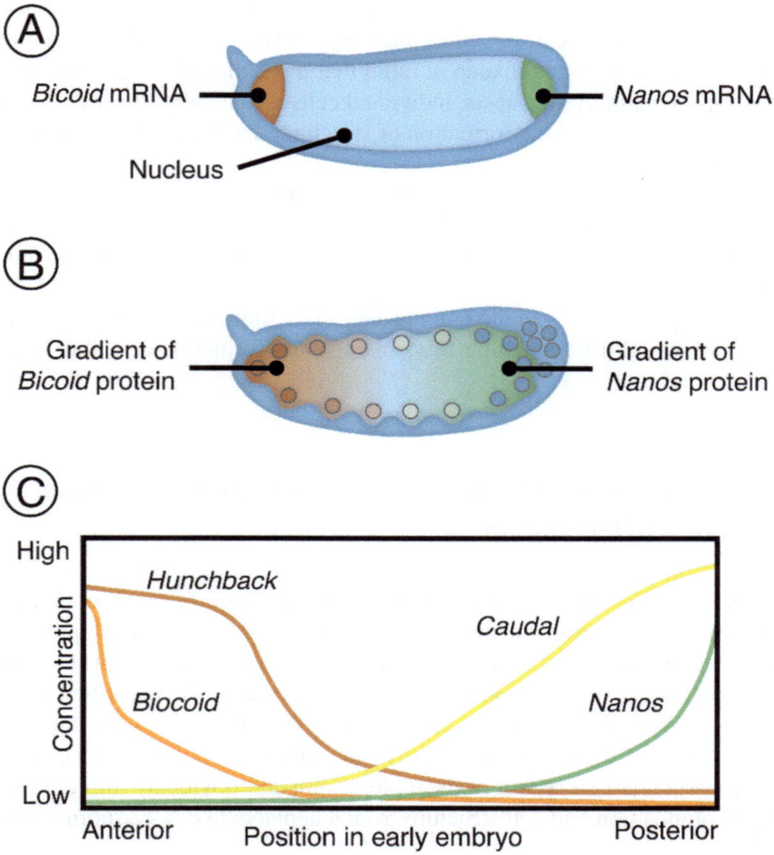

Fig. 3.12 Gradients of different morphogens in *Drosophila* embryos. Bicoid and Hunchback, regulate the production of anterior structures, while, combinations of Nanos and Caudal regulate the formation of the posterior parts of the embryo [2, 5, 11, 14]

agent to suppress the reaction and an excitatory agent to activate the reaction (Fig. 3.12 explains the mechanism of morphogenesis by diffusion chemical gradients) [5, 14, 29]. The accuracy of intercellular reaction-diffusion theory later was proved by several biologists through the polarity formation effect of chemical gradients [12, 14, 25]. Figure 3.12 is a schematic image that illustrates the mechanism of polarity formation at early stages of embryonic development in *Drosophila* [5, 14].

The formation of anterior-posterior polarity of the body is instructed by the anterior-posterior distribution of morphogens in the egg. The initial morphogens may originate from mother's ovaries. Gradients of maternal mRNAs encode transcriptional and translational regulatory proteins that diffuse through the syncytial blastoderm and activate or repress the expression of certain zygotic genes. Appropriate gradient concentrations of Bicoid and Hunchback, regulate the production of anterior structures, while, Nanos and Caudal, regulates the formation of the posterior parts of the embryo [2, 5, 11, 14].

In summary gradient concentration morphogens leads to the dynamic rearrangement of cells which can be observed through the individual and collective cell migrations. Furthermore, we should keep in mind that more than physical rearrangement of cells, during the morphogenesis, individual cells also undergo internal biochemical changes which lead to the formation of structural and functional diversity in the initial stem cells' population.

These changes in cellular biochemistry and function are operated by reciprocal interactions between the inter and extra cellular molecular circuits. As discussed earlier most of computational functions of cells are operated by protein signaling circuits.

In the next part we refer to the main protein signaling pathways that are involved in specification and differentiation of stem cells and formation of embryonic germ layers.

3.5.2 Developmental Biology of Autonomous Specification and Differentiation

As discussed above, morphogen gradients instruct the specific arrangement of stem cells and pattern formation during the early stages of embryonic development. After the fertilization, cell growth starts with a homogeneous population but gradually shifts to the formation of heterogeneous cells which can be distinguished from each other by biochemical and physical properties. Differentiation is only the last overt stage of a series of events that commit the developing population of stem cells to become a particular cell type. For example, a skin cell obviously differs in its biochemical composition and cell structure from a hepatocyte or a cardiomyocyte [15, 21, 27].

The process of commitment of stem cells can be divided into two stages including specification and determination. The specification is a process where stem cells are able to be specified autonomously into different lineages when they are placed in a neutral environment such as a petri dish. At this stage, differentiation potential of stem cells is reversible.

The second stage is determination. A cell or tissue possesses the differentiation potential, even when placed into a non-neutral environment (e.g., in another region of the embryo). At this stage, differentiation potential of stem cells is nonreversible [5]. Specification and differentiation of stem cell during germination leads to the generation of three germ layers. The further commitment of cells at each germ layer causes the formation of all different types and tissues in the body (Fig. 3.12). Specification, differentiation, and commitment of stem cells occur through the activation of highly specific cell signaling pathways toward each cell lineage [5].

In addition to the internal program of cell signaling pathways, stem cells' behavior is affected by environmental factors. In fact, stem cells can understand and adapt themselves to the environmental conditions [15, 21, 27].

This means the internal molecular networks are flexible enough to be adapted with dynamic extrinsic signaling stimuli, as well. In fact, stem cells have a unique endogenous cognitive and decision-maker coding system that is able to guide them towards different destinations based on the dynamic interactions with environmental factors.

Below, we will discuss the main genetic and protein signaling networks that regulate early stages of cell lineage specification (from stemness and pluripotency toward the commitment into three germ layers).

Pluripotency is a distinguishing characteristics of stem cells which provides the differentiation potential towards three main embryonic germ layers. Results of in vitro and in vivo studies on initial population of embryonic stem cells reveals that Oct-4, Sox-2, and Nanog are the key regulatory factors of pluripotency circuitry [21]. Oct-4 is an essential gene regulatory factor which is expressed at the pre-implantation stage. Oct-4 deficient embryos can survive until the morula stage, but, cannot enter the blastocyst stage in both in vitro and in vivo conditions. Oct-4 binds to the octamer motif (5′-ATGCAAAT-3′) of DNA to control the expression of a cluster of genes involved in pluripotency. For instance, there is a close correlation between the expression of Oct-4 and Sox-2 [27].

Sox-2 belongs to Sox family of transcription factors that have a highly conserved DNA-binding domain. Sox2 expression is widely distributed in different tissues of a developing embryo, including ICM, epiblast, neural tissues, and extra-embryonic ectoderm [21]. Sox-2, is an essential factor for ESC pluripotency, as knockdown or conditional deletion of Sox-2 results in trophoblast differentiation. This phenotype is similar to that caused by Oct-4 deletion because Sox-2 often acts as a heterodimer with Oct-4 to regulate transcription of important down-stream genes such as Fgf-4, Nanog, Lefty1.

Nanog is a homodomain-containing protein which functions in coordination with Oct-4 and Sox-2. Genome-wide mapping of Nanog binding sites has identified many pluripotency genes, including Esrrb, Rif1, Foxd-3 and REST [21]. Importantly, many of the self-renewal and pluripotency factors work in cooperation with each other to maintain pluripotency. For example, Sall-4 and Esrrb have been shown to interact with Nanog physically and co-occupy Nanog genomic sites in mouse ESCs [5, 21, 26]. Results of developmental biology studies shows that SMAD1 and STAT3-binding sites are associated with the Oct-4, Sox-2, and Nanog core factor-binding loci. They also share many common regulatory coordinators such as Klf-4, Esrrb, c-myc.

Given that mouse ESCs can be maintained under LIF/BMP condition that enables SMAD1 and STAT-3 activation and binding to genomic sites, this observation provided direct evidence that LIF/BMP signaling supports self-renewal by strengthening core pluripotency circuitry [5, 21]. In the mammalian embryo, the transition from blastocyst to gastrula is a remarkably elaborate process involving a complex series of molecular and cellular events. In addition, expression of Oct-4 and Sox-2 together regulates the cell proliferation though the activation of ERK/MAPK signaling pathway. Furthermore, during preimplantation, successive cleavages of the zygote lead to polarity formation in the body of growing embryo. At this stage blastocyst cavity is surrounded by the trophectoderm (TE). Subsequently, the primitive endoderm (PrE) becomes apparent at the surface of the ICM in contact with the blastocyst cavity.

Transition from pluripotency to lineage specification occurs smoothly by the mechanism of gradient concentration and different combinatorial arrangement of pluripotency signaling factors with the spatiotemporal environmental factors. This leads the activation of downstream lineage specification pathways. For example, high expression level of Sox-2 and Sox-1 is associated with neuroectoderm specification. While high levels of Oct-4 and Nanog along with the up-regulation of Brachyury leads to the mesoderm lineage specification. Wnt/ b-catenin pathway, also plays a central role in differentiation of ESCs towards mesoderm lineage [5]. Lineage specification markers such as Fgf-5 and Cer-1 are expressed in the epiblast, but are absent in the ICM [5, 21]. In parallel to activation of lineage specification and differentiation pathways, developing stem cells, gradually change their position and communicate with new neighbors which leads the establishment of multilayer body plans. Collective migration and rearrangement of cells during the gastrulation lead the formation of three germ layers including ectoderm, mesoderm and endoderm and eventually formation of all different tissues and cell types in the organism's body [5, 7, 21–23, 26].

Here our major goal has been to specifically discuss the mechanism of diversity formation during organogenesis and how combinatorial arrangement of genetic and protein networks leads the diversity formation from the initially homologous population of embryonic stem cells. Detailed description of the whole embryonic developmental pathways towards the end stage differentiation of various tissues has been out of scope of this chapter. For detailed information, we highly recommend reviewing the relevant citations in the list of references [5, 21, 24, 26, 27, 30].

In overall, we can summarize the main properties of embryonic organogenesis as following:

1. Self-organization is a bottom-up process which starts from the molecular scale and can progress freely through multiple levels of complexity.
2. The process is precisely programmed by an internal molecular algorithm which is also responsive to the physiochemical environmental factors (e.g. through the electrochemical attractive forces between molecules and atoms).
3. The reciprocal interactions of intrinsic cell signaling networks with each other and environmental factors makes a very dynamic operating system for the organism.
4. The process is operated through a dynamic and autonomous algorithm where the output of each step functions as an input for the next stage.
5. The process is accompanied by autonomous and progressive pattern formation.

By considering the above mentioned properties of self-organization, it can be concluded that self-organization in biological systems functions similar to a Mandelbrot algorithm, where the output of each stage is an input for the next stage. Therefore, as long as the operation is running, the algorithm can generate more sequences of dynamic patterns. Also, the algorithm is able to transfer freely among multiple layers of complexity.

Considering the molecular-based property of biological operating systems, it can be concluded; a critical requirement of self-organization is related to the unique capacity of information storage and processing by biomolecules. The tri-fold functional property of molecules also can be translated as relativity in code, energy, and mass for detailed description of different aspects of self-organization. Relativity of code, energy, and mass is a distinguishing property of life, which differentiates a living system, from any other non-living elements of nature.

3.5.3 Molecular and Physical Foundation of Embryonic Self-Organization by Cognitive Chemistry and Relativity of Code Energy and Mass

Can we simulate the developmental mechanisms of self-organization beyond biology?

Scientifically, by considering the chemical foundation of biological systems, it is possible to simulate the self-organizing property of embryonic stem cells. However, the level of complexity of such systems depends on deep understanding of the molecular operation system of a living organism and how basic principles of physics and chemistry are applied in designing of living systems by nature. Same basic principles can be applied in generation of similar synthetic systems. In the next part of this chapter and also chapter 6, we try to explain the chemical foundation of self-organization through the basic principles of cognitive chemistry. We will explain the properties of optimal materials for generation of self-organizing systems in synthetic forms.

Chemical foundation and molecular properties of self-organizing systems (major characteristics of optimal materials for generation of synthetic self-organizing systems):

1. Internally coded materials and molecules: instruction for the final 3D conformation of a geometrical structure needs to be embedded in the electrochemical attractive forces between atoms of a molecule.

Figure 3.13a is an schematic representative of the information stored in the sequence of amino acids. Figure 3.13b illustrates the flexible structure of protein conformation. Unlike any other crystals in nature (with static 3D structure), proteins have highly flexible 3D structures and based on the environmental conditions can obtain different conformational structure. In other words, proteins have a multidimensional geometrical structure in response to the environmental factors. This unique chemical property of proteins leads to their multifunctional properties including molecular signal transducers, logic gate switches, molecular machinery, molecular sensors, etc.

2. High flexibility in the 3D structure of proteins can be defined as the main molecular mechanism of recognition, memory formation, and adaptation with environmental factors in living cells. Dynamic changes in the conformational structure

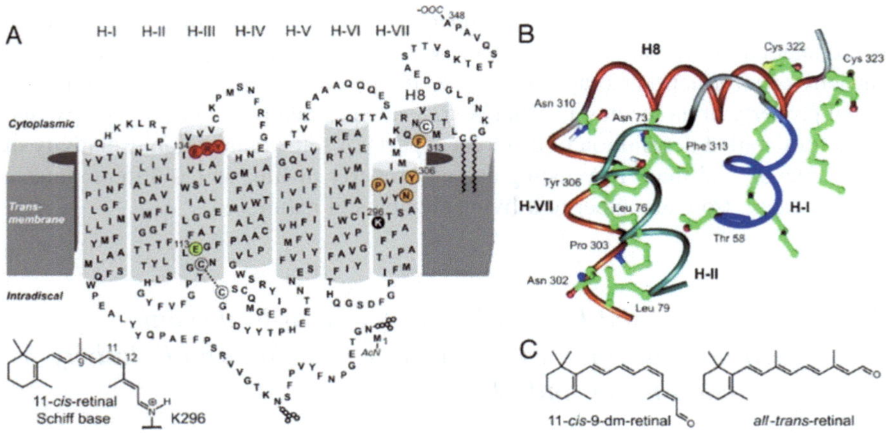

Fig. 3.13 (**a–c**) The schematic image represents the multifunctional capacity of proteins (coding units, structural and machinery functions, at the same time). The schematic represents the crystallography of rhodopsin structure. The instruction for the 3D structure of the protein is embedded in the sequences of amino acids, which also is critical in its machinery functionality. (Image adapted from Fritze et al. [4])

of proteins can be observed in molecular sensors, receptors, molecular logic gates, and signal transducers based on environmental factors. Combinatorial interactions of these molecules generate different cell-signaling pathways.

DNA has been the major focus of studies in designing biomimetic coding systems. However, proteins also are promising models for the generation of synthetic coded materials. Proteins have multifunctional capacity due to their special structural properties and are more efficient models for environmental recognition and dynamic memory formation.

Furthermore, proteins are the best models for the generation of self-organizing polymers. All instructions related to the final 3D structure of a protein are embedded in the electrochemical interactions among the side chains of amino acids' sequences. In fact, electrochemical interactions among the side chains of amino acids generate a coding system which regulates the dynamic conformational structure of a protein based on the variations in the environmental conditions. Therefore, the coding system of amino acids instructs the dynamic or 4D structure of a protein, based on environmental factors (signals).

Figure 3.13 illustrates a schematic image representation of the flexibility in the conformational structure of rhodopsin, which is a light-sensitive receptor protein involved in visual phototransduction. Rhodopsin belongs to the category of G-protein-coupled receptors (GPCR) and is found in the rods of retina.

Rhodopsin is extremely sensitive to the light. The 3D conformational structure of rhodopsin changes immediately after exposure to the light. Dynamic changes in the conformational structure of rhodopsin consequently lead to the variation in the affinity of this protein to the other intercellular signaling molecules [4]. This process finally is identified as a biological signal transduction. In fact, unlike any other non-living crystals (with 3D structure) in nature, proteins illustrate 4D conformational structure.

This unique chemical-based coding system, which is embedded in the sequences of amino acids, is a remarkable property that distinguishes the 4D and dynamic structure of proteins from any other static crystals in nature. The unique coding system of amino acid sequences and the 4D (dynamic) conformational structure of proteins make them function as highly efficient signal transducers and logic gates at the same time.

In fact, each protein functions as a separate computer that is able to store information, not only in the sequences of their subunits but also in their 3D structure. Protein signal transducers are able to convert different types of physico-chemical environmental signals into the computable signals for the cells. The multifunctional property of proteins causes a very fast reaction with the environmental signals.

3. In addition, data processing in signaling protein is based on the spontaneous electrochemical interaction of atoms in the amino acids' side chains. Therefore, solving complex problems in the interface with the environmental factors (e.g., NP problems) occurs in real time.

Another example of information stored in the 4D patterns of proteins can be explained by the mechanisms of memory formation in the patterns of brain neural networks and through inter- and intracellular protein and electrochemical signaling pathways among neural cells. Information storage in the 4D patterning of neural cells and protein-protein signaling pathways causes the expansion of memory capacity of neural cells. On the other hand, 4D patterning of proteins during embryonic development is accompanied by cell cleavage that causes the physical growth of an embryo.

It is possible that during the activation of neural cells, combinations of different types of neurotransmitters influence the structures of signaling proteins in neural cells. This leads to the formation of completely new structures and activation of completely new sets of signaling pathways. Addition of these pathways causes the expansion of memory capacity of neural cells. Therefore, unlike any other human-made computers with limited information storage capacity, neural cells have expandable memory capacity, meaning the brain memory capacity is expandable as the input information increases (e.g., the brain's information storage capacity is enhanced by education).

References

1. Alberts B, Johnson A, Lewis J, Raff M, Roberts K, Walter P (2008) Molecular biology of the cell, 5th edn. Garland Science Taylor and Francis Group, New York
2. Christian JL (2012) Morphogen gradients in development from form to function. Wiley Interdiscip Rev Dev Biol 1(1):3–15
3. Deglincerti A et al (2016) Self-organization of spatial patterning in human embryonic stem cells. Curr Top Devel Biol 116:99–113
4. Fritze O et al (2003) Role of conserved NP xxY(X) 5,6,F motif in the rhodopsin ground state and during activation. PNS 100(5):2290–2295
5. Gilbert S (2010) Developmental biology, 9th edn. Sinauer Associates Inc, Sunderland, England

6. Greggio C et al (2015) In vitro- produced pancreas organogenesis models in three dimensions; self-organization from few stem cells or progenitors. Stem Cells 33(1):8–14
7. Huang G et al (2015) Molecular basis of embryonic stem cell self-renewal: from signaling pathway to pluripotency. Cell Mol Life Sci 72(9):1741–1757
8. Kauffman SA (1993) The origins of order: self-organization and selection in evolution. Oxford University Press, New York
9. Kauffman SA (1995) At home in the universe: the search for the laws of self-organization and complexity. Oxford University Press, New York
10. Kauffman SA (2000) Investigations. Oxford University Press, New York
11. Kwan CW et al (2016) Functional evolution of a morphogenetic gradient. eLife 5:1–12
12. Lauschke VM et al (2013) Scaling of embryonic patterning based on phase gradient encoding. Nature Letter 493:101–105
13. Lee SS, Shibata T (2015) Self-organization and advective transport in the cell polarity formation for asymmetric cell division. J Theor Biol 382:1–14
14. Lewis EB (1978) A gene complex controlling segmentation in Drosophila. Nature 276(7):565–570
15. Mark F, et al (2009) Cellular signal processing; an introduction to the molecular mechanism of signal transduction. Garland Science Tylor and Francis Group, New York, USA
16. Marsh JA, Forman-Kay JD (2010) Sequence determinants of compaction in intrinsically disordered proteins. Biophys J 98:2383–2390
17. Moret MA (2011) A self-organized critical model for protein folding. Physica A 390:3055–3059
18. Nelson DL, Cox MM (2017) Lehninger principles of biochemistry, 7th edn. Freeman, W H& Company, New York
19. Nepusz T, Vicsek T (2013) Hierarchical self-organization of non-cooperating individuals. PLoS One 8(12):1–9
20. Northrop B, Zheng Y, Chi K, Stang P (2009) Self-organization in coordination-driven self-assembly. Acc Chem Res 42(10):1554–1563
21. Oron E, Ivanova N (2012) Cell fate regulation in early mammalian development. Phys Biol 9:1–19
22. Partiff DE, Shen MM (2014) From blastocyst to gastrula: gene regulatory networks of embryonic stem cells and early mouse embryogenesis. Philos Trans R Soc Lond B Biol Sci 369(1657):1–12
23. Rodwell VW, Bender D, Botham K, Kennelly P, Weil PA (2015) Harpers illustrated biochemistry, 30th edn. The McGraw Hill Education, New York/London
24. Rossant J, Tam PP (2009) Blastocyst lineage formation, early embryonic asymmetries and axis patterning in the mouse. Development 136(5):701–713
25. Sanger A, Briscoe J (2017) Morphogen interpretation, concentration, time, competence, and signaling dynamics. Dev Biol 6:1–19
26. Sasai Y, Eiraku M, Suga H (2012) Invitro organogenesis in three dimensions: self-organizing stem cells. Development 139:4111–4121
27. Shahbazi MN et al (2016) Self-organization of the human embryo in the absence of maternal tissues. Nat Cell Biol 18(6):700–708
28. Simister PC et al (2011) Self-organization and regulation of intrinsically disordered proteins with folded N-termini. PLoS Biol 9(2):1–4
29. Turing AM (1952) The chemical basis of morphogenesis. Philosophical Transactions of the Royal Society of London Series B Biological Sciences 237(641):37–72
30. Wartlick O et al (2014) Growth control by a moving morphogen gradient during Drosophila eye development. Biol Develop 141:1884–1893
31. Zeeuo T, Weijers D (2016) Plants organogenesis: rules of order. Curr Biol 26:157–179
32. Zernicka-Goetz M (2002) Patterning of the embryo: the first spatial decisions in the life of a mouse. Development 129(4):815–829.

Chapter 4
Molecular Mechanism of Self-Fueling in Biological Systems by Relativity of Code, Energy, and Mass

Abstract Among the renewable energy resources, solar energy is the largest. The increase in the use of fossil fuels and increasing level of CO_2 lead to an emerging demand for sustainable energy sources. Nature applies a highly efficient way to harvest solar energy. Photosynthesis is a natural process through which solar energy is absorbed by plants and transformed and stored as chemical energy in biomolecules. During photosynthesis, photochemical reactions happen in light-harvesting complexes in chloroplasts of plants' cells.

In addition, photosynthesis is incorporated with the formation of highly organized structures from the less ordered materials (self-organization). High efficiency of photosynthesis is related to the quantum coherence energy transfer within light-harvesting protein complexes.

Description of the mechanism of energy and mass transfer in photosynthesis relies on the multifunctional properties of biomaterials at different levels of complexity including subatomic (quantum), atomic, and molecular levels. The multifunctional property of matter for energy, mass, and information transport at subatomic, atomic, and molecular levels makes up the chemistry of cell, as a very flexible matrix for free transportation of energy, mass, and information.

Keywords Self-fueling systems · Natural and artificial photosynthesis · Renewable and sustainable energy solutions · Quantum coherence energy and information transfer in biological systems

4.1 Self-Fueling Property of Living Systems and the Maintenance of Cell Organization

Maintenance of information enriched environment and structural organization of living cells require continuous energy input. Disorder in energy input to a cell results in dissociation of its organization or cell-death [19, 23, 27].

Living cells obtain their energy requirements from the oxidation of carbon-enriched organic biomolecules. This process which is called respiration is accompanied by the production of carbon dioxide and water [20]. Oxidation of carbon-enriched

T. Karimi, *Molecular Mechanisms of Autonomy in Biological Systems*,
https://doi.org/10.1007/978-3-319-91824-2_4

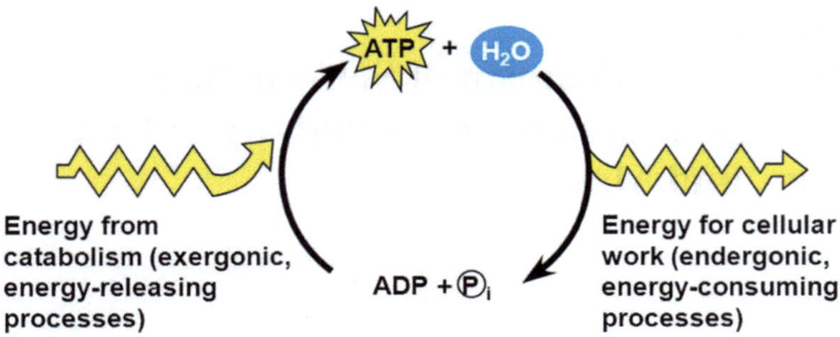

Fig. 4.1 Schematic image representative of coupling of exergonic and endergonic reactions among catabolic and anabolic pathways in living cells

macromolecules in a living cell (through respiration) produces energy. However, the output energy of respiration never releases in the cell directly as heat. However, in the cell, each exergonic reaction of respiration is coupled with an endergonic biosynthesis reaction [1, 27]. Coupling of respiration and biosynthesis reactions makes a very efficient energy transfer system via the metabolic pathways. Oxidation of biomolecules in living cells is regulated in the cell through catabolic pathways, while anabolic pathways play a role in de novo synthesis of biomolecules in the cell (Fig. 4.1). Carefully designed coupling of catabolic and anabolic pathways is a highly efficient way for energy transition in the cell and causes the autonomous factory-like behavior of cell (Fig. 4.2).

The dedicated coupling of a sequence of catabolic and anabolic pathways in the cell also causes the path-dependence progress of metabolic pathways [19, 27]. Sequences of coupled chemical reactions can function as a domino where flow of energy transfers freely among the sequences of reactions, one after the next one. In fact, by this mechanism, carefully coupled sequences of chemical operations can program the chemistry of cell to function as a large molecular domino. When a metabolic pathway activates via the initial reaction, the process continues through the sequence of reactions in a path-dependence manner. The sequential progress of biochemical reactions in metabolic pathways causes the autonomous and factory-like behavior of living cells (e.g., the autonomous progress of biosynthesis reactions).

According to the second law of thermodynamics, the degree of disorder (or entropy) increases in the universe, spontaneously [1, 27]. In other words, a system progresses spontaneously toward the greater entropy. In biological systems, all chemical reactions progress toward the formation of more organized structures. Therefore, living cells require energy to maintain their highly ordered organization. On the other side, living cells perform all operations autonomously. In fact, living cells also are able to obtain their energy requirements from their environment, autonomously. This special capacity of the living system for autonomous uptake of energy from their environment is also defined as self-fueling.

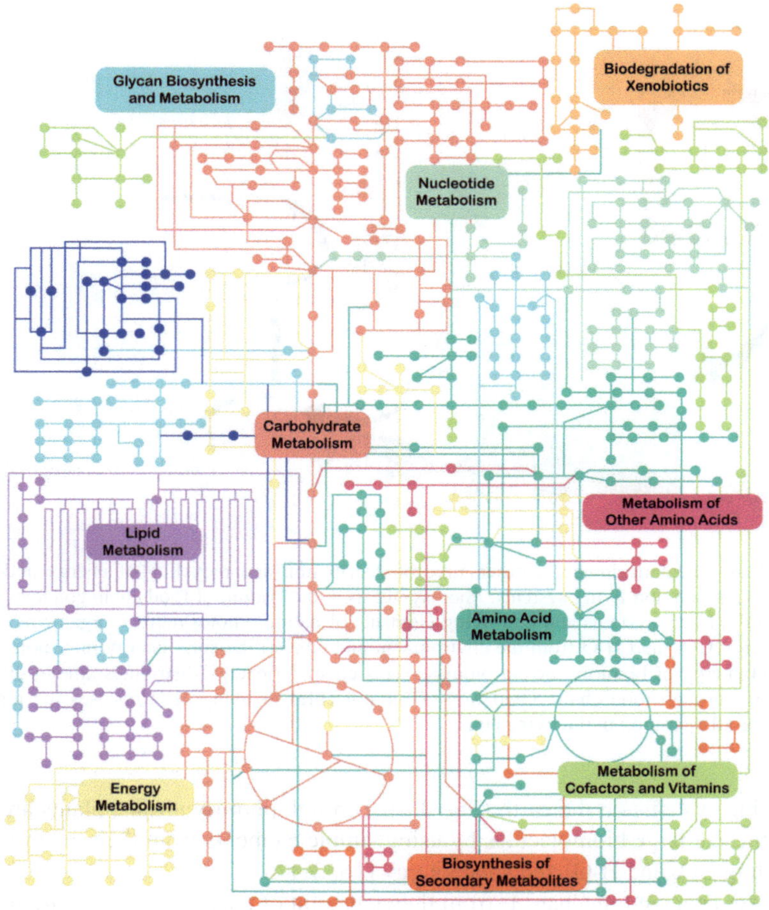

Fig. 4.2 Schematic image representative of a cell metabolic map. Schematic illustrates the tight connections between the anabolic and catabolic pathways in four main groups of biomolecules including carbohydrates, proteins, nucleic acids, and lipids. (Image from http://manet.illinois.edu/pathways.php)

The self-fueling capacity of living cells is regulated by a set of precisely programmed oxidation-reduction reactions. Unlike any other non-living system, the flow of energy in living cells is designed as a carefully programmed process. However, still, energy and mass transfer freely among living and nonliving systems in the nature. That means living cells are open systems (by the free transfer of energy and mass with their environment) [1, 23].

Biological systems obtain their energy requirements from the chemical bonds of organic molecules or directly from sunlight. At the bottom of the animal food chain, herbivores obtains energy from plants. Plants absorb energy directly from sunlight through photosynthesis. In fact, photosynthesis provides a link between non-living

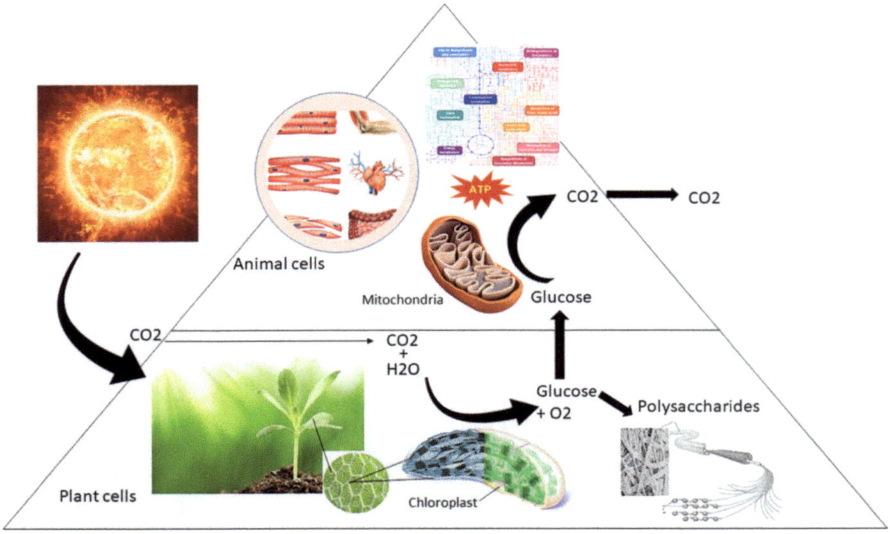

Fig. 4.3 A schematic image representative of distribution and recycling of energy and mass among living and non-living systems. Photosynthesis at the bottom of food chain connects a link between the living systems and non-living elements of nature. Photosynthetic organisms absorb solar energy CO_2, and H_2O from the environment and convert the solar energy into more stable chemical energy of biomolecules in biomass. This chemical energy distributes among different organisms through food chain. Oxidation of biomolecules in living cells releases the required energy for all operation in the living cells

and living systems on the earth (Fig. 4.3). Through photosynthesis, plants are able to synthesize large biomolecules by using simple elements from their environment (including CO_2 and H_2O) and solar energy [16, 21].

As we discussed in Chap. 1, from the viewpoint of energy and mass circulation, living systems are highly connected to each other and their environment (Fig. 4.2). Plants, animals, microorganisms, and non-living elements of nature have existed on this planet for billions of years, adapted together, and become an essential part of the others [1, 23, 29]. Figure 4.2 illustrates a schematic image representative of circulation of energy and mass among living and non-living elements of nature.

Current understanding of nature's laws in physics and chemistry is limited to the relativity of energy and mass. The relativity of energy and mass can explain physicochemical properties of non-living elements of nature. However, still, there are several features in the nature that cannot be explained simply if we consider energy and mass as the only aspects of the matter. In this chapter, we discussed the self-fueling capacity of the biological system as a precisely programmed energy transfer system. Self-fueling property of a living organism is precisely ptogrammed by a set of chemical reactions which provide energy from environmental resources for the organism. In this chapter, photosynthesis is studied as a molecular machinery and a linker between the sun and formation of life from non-living elements of

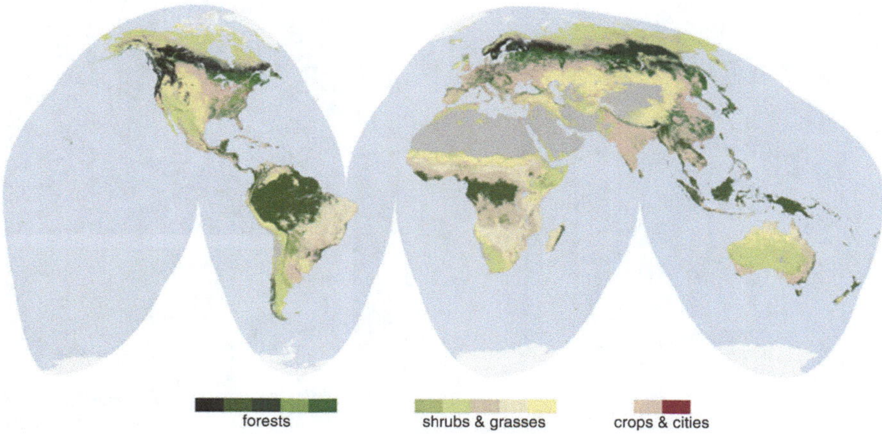

Fig. 4.4 Distribution pattern of photosynthesis organisms on the earth. Forests cover 30% of earth (dark green areas on the map are related to taller forests). Areas related to the shrubs, grasses, and smaller photosynthetic organisms are defined by light green color. (Image courtesy of NASA, Earth observatory resource, http://earthobservatory.nasa.gov/Features/ForestCarbon/)

nature. Deep understanding of mechanisms of uptake and transfer of solar energy into biomass can unmask several unknown aspects in the borderline between living systems and non-living elements of nature. In addition, photosynthesis is a promising natural model for future sustainable energy solutions.

4.2 Photosynthesis: A Natural Sustainable Energy Solution for Life

Photosynthesis is a natural process by which solar energy converts to chemical energy through a cluster of carefully programmed photochemical reactions in photosynthetic organisms (including plants, algae, and purple bacteria). Photosynthesis provides the base energy requirements for sustaining of almost all life on earth [3, 23, 24, 29]. Forests cover about 30% of the earth's land (Fig. 4.4) [5, 22].

Through natural photosynthesis, plants absorb sunlight to split molecules of water into hydrogen and oxygen [21]. Oxygen releases to the atmosphere and would be used for maintenance of living cells. Hydrogen would be used to reduce and recycle carbon dioxide from the atmosphere as well as production of biomass [3, 17, 22]. Furthermore, a unique feature of photosynthesis is the autonomous generation of designer mass by using less ordered materials (including water, CO_2 from atmosphere) and solar energy (Fig. 4.5).

Deep investigation of photosynthesis in green plants is critical for clearer descriptions of basic principles of physics and chemistry in biological systems. Same principles can be applied for future in-situ resource utilization as well as sustainable energy solutions.

Fig. 4.5 Autonomous formation of designer mass through photochemical reactions of photosynthesis by using solar energy and less ordered materials (CO_2 and H_2O)

In the next part, specifically we will discuss the regulatory mechanisms behind photosynthesis reactions including (1) mechanisms of light absorption through quantum coherence effect in light-harvesting complexes, (2) conversion and storage of solar energy into the chemical bond of biomolecules, and (3) autonomous generation of designer mass by self-organization of biomolecules.

4.3 Quantum Coherence: The Efficiency of Solar Energy Uptake and Conversion in Photosynthesis

Photosynthesis is a process by which plants absorb the energy from sunlight, carbon dioxide, and water to produce biomass. In fact, photosynthesis makes a critical link between the energy of the sun and different forms of life on the earth.

Figures 4.6 and 4.7 illustrate the structure of two categories of photosynthetic organisms including primitive unicellular photosynthetic bacteria and multicellular algae. Figure 4.8 illustrates a schematic image which compares the main functional units of photosynthesis in plants' chloroplasts with Cyanobacteria.

Photosynthesis reactions take place in two main stages, including (1) light-dependent reactions and (2) light-independent reactions. Figure 4.9 illustrates a schematic image representative of different steps of photochemical reactions in a plant's chloroplast.

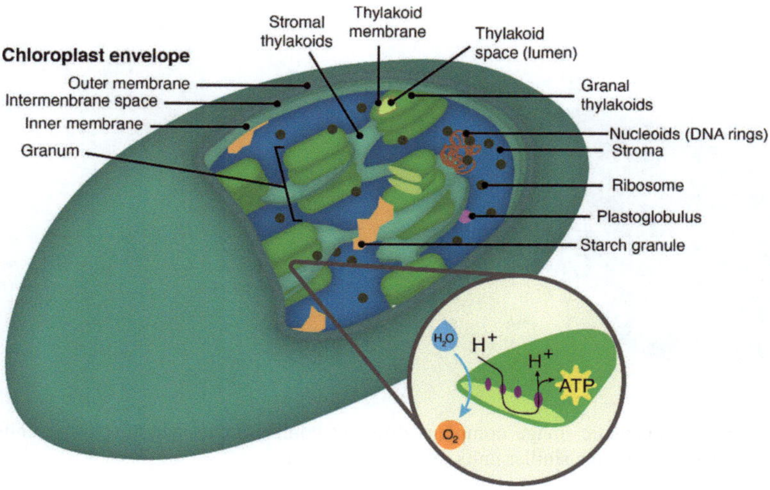

Fig. 4.6 A schematic image representative of different compartments of a chloroplast

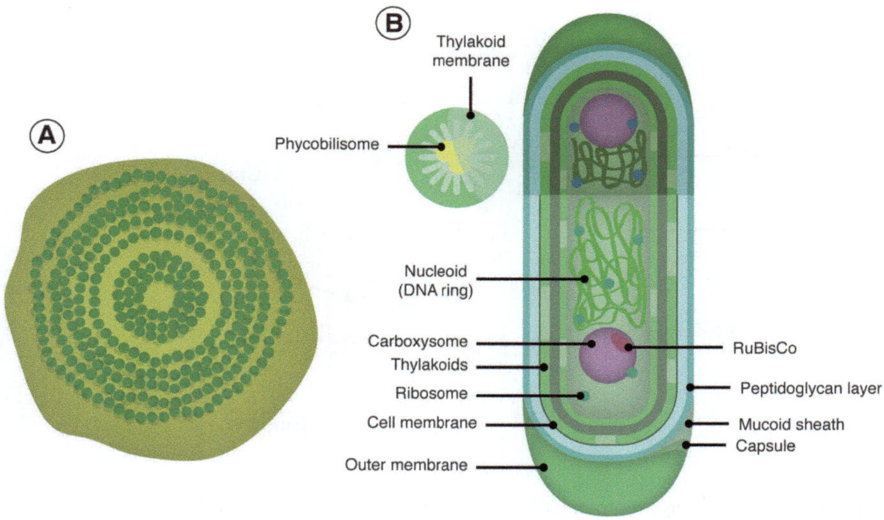

Fig. 4.7 (**a**) Microscopic image filamentous *Cyanobacteria*. (**b**) Schematic image representative of different parts of *Cyanobacteria*. (Images are adapted from https://en.wikipedia.org/wiki/Cyanobacteria)

In the first stage of photosynthesis, the energy of photons from sunlight is absorbed in the light-harvesting protein complexes of thylakoid's membrane, while in the second step this energy stores in the form of more stable chemical energy (e.g. chemical energy of carbohydrates).

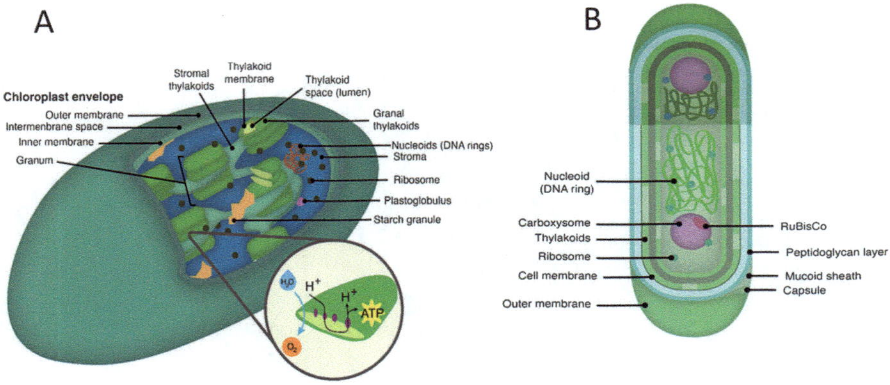

Fig. 4.8 (**a**, **b**) Schematic image compares different compartments and functional units in a plant's choloroplast with the similar units in a cyanobacteria

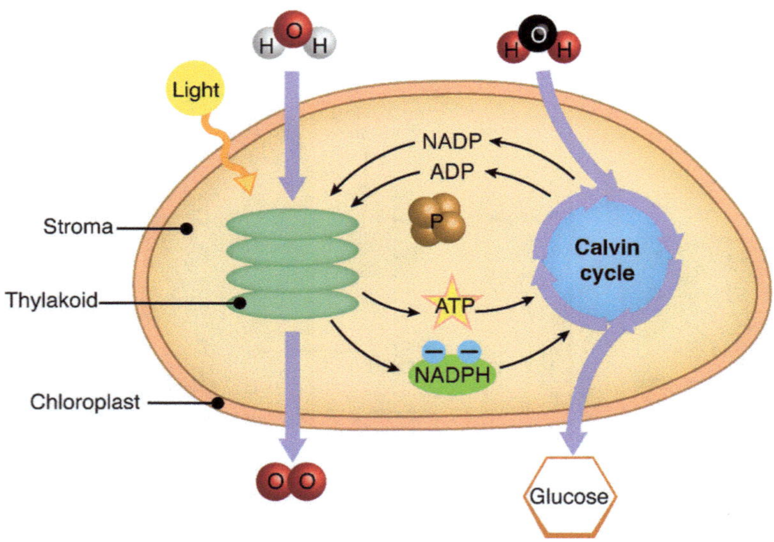

Fig. 4.9 Schematic image representative of different steps of photochemical reactions in a plant's chloroplast including (1) light-harvesting reaction in the thylakoid membrane and (2) Calvin cycle and production of organic molecules in the chloroplast stroma

Figure 4.10 illustrates a schematic image representative of energy transfer between the light-harvesting protein complex and the photosynthesis center II in the thylakoid membrane of chloroplast in a plant's cell. The four main compartments in plants thylakoids are including photosystem II (PSC II), cytochrome b_6f, photosystem I (PSI), and the ATP synthase [12]. Figure 4.10 illustrates a schematic image of the thylakoid membrane and the main photosynthetic protein complexes.

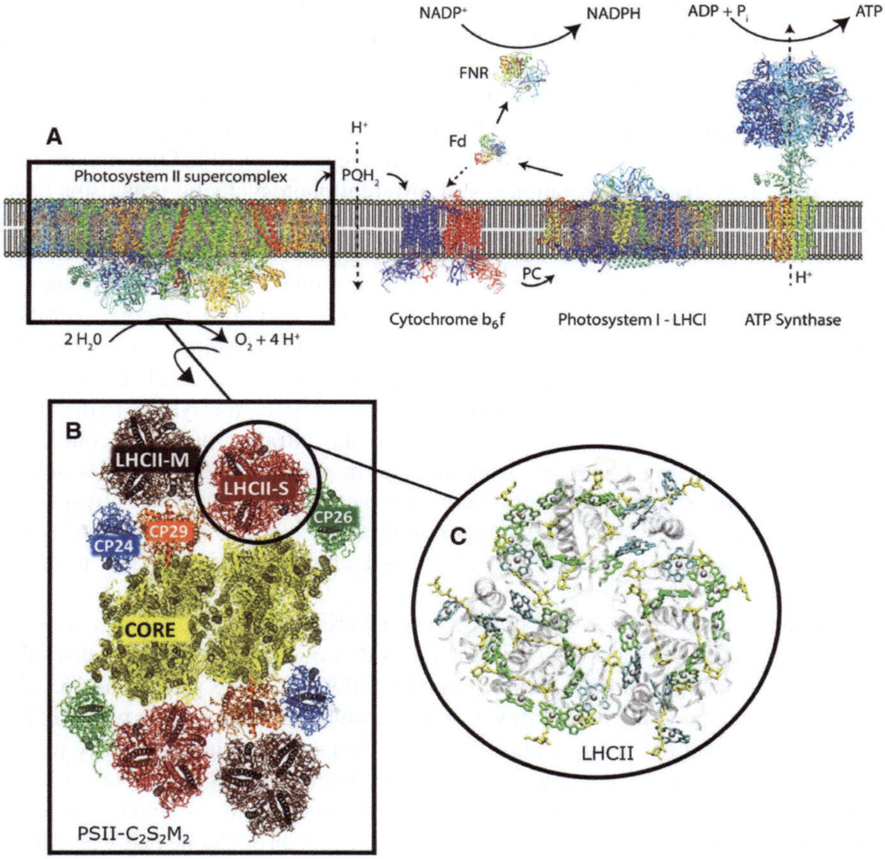

Fig. 4.10 A schematic image representative of the thylakoid membrane and the main photosynthetic protein complexes. (**a**) Schematic image representative of the order of reactions that happen in photosynthesis protein complexes, including photosystem II, cytochrome b_6f, photosystem I (LHCI), and ATP synthase. (**b**) Schematic representative of different subunits of photosystem II (PSC II). (**c**) Closer view of LHCII antenna. (Image courtesy of [12])

The highly efficient absorption and transfer of solar energy is related to the remarkable function of specialized light-harvesting complexes (also called antenna complex). Light-harvesting complex is localized in the thylakoid membrane of chloroplasts. Solar energy in the form of photons absorbs to light-harvesting antennas and causes an electronic excitation [18, 28]. The solar excitation energy transfers from each antenna to a reaction center. In the next step, the charge separation in reaction center transforms electronic excitation energy of photons to the more stable chemical energy [4, 8, 18, 28].

Efficiency of light energy transfer in light-harvesting protein complexes has been the subject of intense research in both biology and physics. Engel et al. [9], reported the direct evidence of quantum coherence energy transfer in light-harvesting apparatus of green sulfur bacteria. Results of Engel et al. [9], revealed that energy of photons was collected and transferred to the reaction center through a specialized light-harvesting structure called the Fenna-Matthews-Olson (FMO) complex [9]. Since then, light-harvesting complexes of different photosynthetic organisms have been studied intensively [8, 12, 13, 18, 26].

Results of several studies reveal that light energy absorbtion and transfer among antenna and photosynthesis center II occurs with remarkable quantum efficiency. The excitation energy typically delocalizes and transfers among different chromophores. These localized excited states are called exciton. Exciton states occur in pigment compartments of antenna which are embedded in a protein bath. The protein bath also facilitates the quantum coherence by allowing the system to dissipate energy [26]. Hildner et al. [13], demonstrated the ultrafast quantum coherence energy transfer among light-harvesting antenna in purple bacteria. Results of this study reveal the existence of quantum coherence energy transfer under physiological condition. Observation of quantum coherence energy transport at room temperature and wet environment of cell still is a challenging subject that needs to be investigated in the future.

One mechanism to explain the ultrafast energy transfer among pigments of light-harvesting complex toward reaction center II is that quantum coherence causes the possibility of energy transfer across multiple pathways at the same time [6, 13]. However, in fact, energy always travels only through the shortest path which is the most efficient way by energy status [13]. This means the excitation energy jumps among light-harvesting antenna by quantum coherence effect. There are a large number of possibilities for energy transfer among antenna (as combinatorial pathways). However, in fact at each moment, energy travels through one pathway which is the optimal way for energy transport. Figure 4.11 illustrates a schematic representative of quantum coherence effect and the possibility of energy transfer via pathways among light-harvesting antenna at the same time. Ultrafast operation of energy transfer in light-harvesting chain is similar to the ultrafast computing capacity of quantum bits in a quantum computer. Unless that, quantum coherence occurs in photosynthesis at physiological temperature.

In the next step, the excitation energy of photons is used to split water molecules. Because of water splitting reaction, proton (H^+), electron, and oxygen are released. Electrons transfer to the LHC to recharge the donated electrons from antenna through their electronic excitation. Oxygen goes to the atmosphere. The release of proton generates a proton gradient that is used to produce energy-carrying molecules (including ATP and NADP) [3, 12].

In the next stage (which is called Calvin cycle), light-independent chemical reactions occur. At this stage, proton-carrier molecules (NADPH and ATP) are used in carbon fixation process which finally leads the de novo synthesis of saccharides (e.g., fructose and glucose) [3].

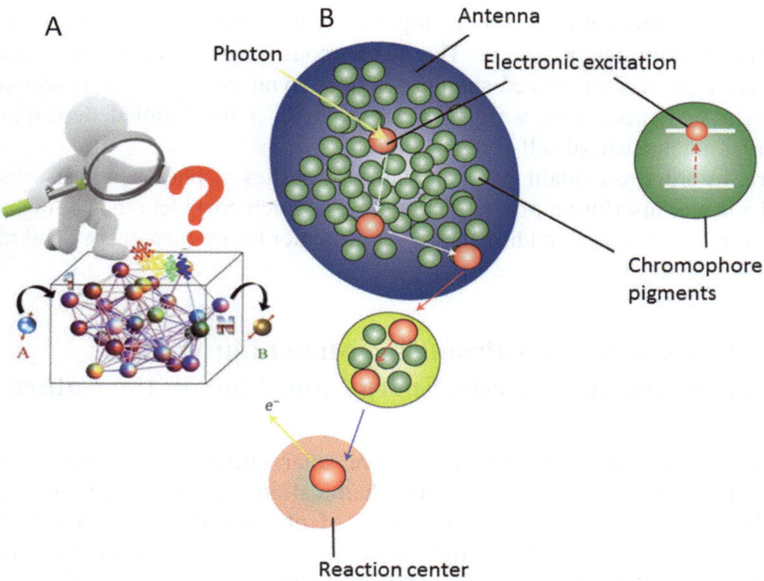

Fig. 4.11 (**a, b**) Schematic image representative of quantum coherence energy transfer in light-harvesting antenna toward photosynthesis reaction center II. Quantum coherence causes the possibility of energy transfer across multiple pathways at the same time. However, in fact, energy always travels only through the shortest path which is the mostly efficient way by energy status. Quantum coherence effect also leads the ultrafast operation of energy transport pathway among all different possible combinatorial pathways. (Part **b** of image adapted from [6])

4.4 Photosynthesis: Quantum Coherence, Energy Transfer, and Production of Designer Mass

Figure 4.5 illustrates different geometrical structures, generated autonomously by photosynthesis in green plants. During photosynthesis, in addition to the energy conversion, the system is able to produce highly organized structures from less organized molecules (CO2 and H2O). Therefore, the photosynthesis process is the combination of conversion of energy from sunlight to chemical energy as well as formation of highly organized molecular structures from less organized molecules (decrease in entropy). Nature is able to produce a designer mass from less organized molecular subunits, by the cost of solar energy.

In other words, photosynthesis is more than conversion of solar energy to the form of chemical energy. Current description of relativity of energy and mass is defined by equation of $E = \text{m}C^2$, only considering energy and mass as the only aspects of matter in the nature.

However, photosynthesis is an endogenously programmed (code-based) energy and mass transformation system. The autonomous formation of highly organized structures from less organized subunits in photosynthesis can be considered as a self-organizing process as well. Please see Chap. 3 for detailed description of molecular mechanism of self-organization.

Here, in order to explain the remarkable properties of photosynthesis, we suggested a new transcription of relativity equation which considers the coding capacity of biomolecules as an additional aspect of matter (more than energy and mass).

4.5 Photosynthesis: A Promising Model Illustrating the Relativity of Code, Energy, and Mass in the Nature

Photosynthesis is a natural process where solar energy is absorbed by light-harvesting complex and converted into chemical energy stored in biomolecules. Absorbed light energy from the first step of photosynthesis (light dependent) transfers to the second phase (light independent) by energy carriers such as NADPH and ATP. In the second step, NADPH and ATP are used to reduce carbon dioxide and generate biomass. In this step, photosynthesis is incorporated into the formation of highly organized structures from the less ordered materials (self-organization). As discussed in Chap. 3, self-organization is an internally programmed process. In other words the instruction for the final structure of biomaterials is embedded in the sequence of their coding subunits. For example, the final structure of cellulose fibers is directly instructed by the sequences of monosaccharides in the polymeric structure of this molecule as well as intermolecular interactions toward the formation of more complex structures at the micro and macro level of complexity.

Overall, it can be concluded that highly ordered transfer of energy and mass in photosynthesis is directly relevant to the multifunctional properties of biomatter in the sense of coding, energy and mass transformation at the same time. Information storage and transfer by biomatter is a special property of biomaterials that has not been explained clearly by the relativity of energy and mass ($E = \mathrm{m}C^2$).

Photosynthesis is a naturally autonomous process that is incorporated with free transfer of energy, information, and mass. The multifunctional property of biomaterials as units of energy, code, and mass also can be observed at different levels of complexity from subatomic level to micro and macro levels of complexity. Therefore, by considering information storage (coding) capacity of biomaterials as the third dimension of matter, here we defined the relativity of code, energy, and mass, to explain the precisely programmed sequences of energy and mass transfer in photosynthesis.

4.6 Photosynthesis Is a Code-Based Energy and Mass Transformation System

As we discussed above, photosynthesis is a code-based energy and mass transformation system. Unlike current solar cells which harvest and convert solar energy to electricity, photosynthesis is accompanied by the careful synthesis of designer materials with highly specific structures (Figs. 4.12, 4.13, and 4.14).

Furthermore, the entire machinery for harvesting and transformation of solar energy is synthesized on demand by plants. Nature applies the most efficient way to harvest solar energy. A copy of all required information for generation of photosynthetic machinery is stored in seeds as an information-enriched matrix. In fact, unlike any other human-made systems, photosynthesis is a code-based energy transformation system that applies information-enriched materials (DNA) to produce a solar energy-harvesting system. Then, this energy-harvesting system applies inherited molecular codes (in the DNA of an organism) to generate designer materials with highly specific physicochemical properties (Fig. 4.13).

In addition, through photosynthesis, plants harvest solar energy and store this energy in activated carriers of energy which function as molecular rechargeable batteries. ATP and NADH are the most important activated energy carriers that are involved in photosynthesis [1, 3, 23]. The net result of the entire process of photosynthesis can be summarized in the equation:

$$\text{Light energy} + CO_2 + H_2O \rightarrow \text{sugars} + O_2 + \text{heat energy}.$$

The produced sugar is then used both as a source of chemical bond energy and as a source of materials to make the many other small and large organic molecules that are essential to the plant cell (Figs. 4.14a, b and 4.15).

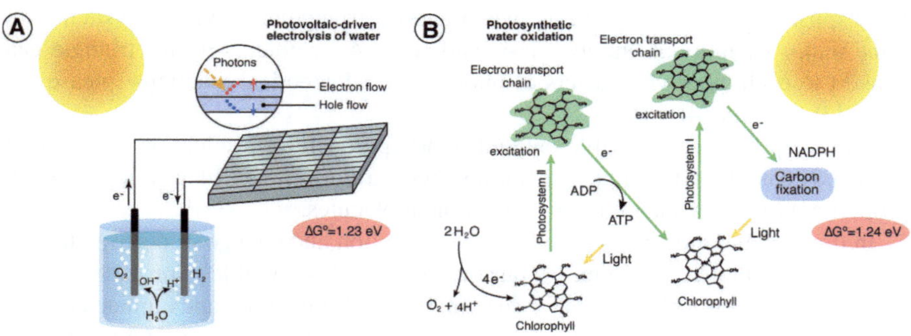

Fig. 4.12 A schematic image compares a current human-made solar energy-harvesting system (which transforms solar energy into electricity – part A) with the photochemical reactions in plants (which leads to the transformation of solar energy to chemical energy and synthesis of biomass)

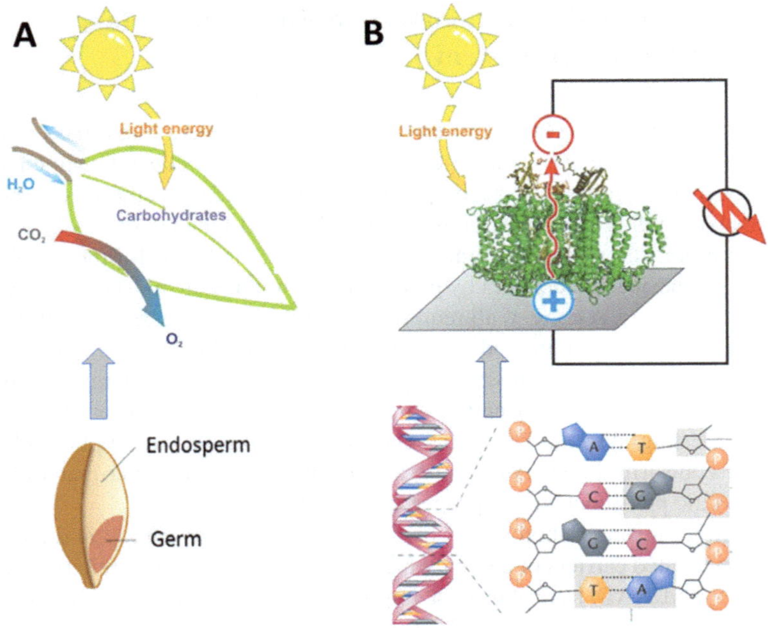

Fig. 4.13 (**a**, **b**) Code-based energy sustainability for generation of machinery

4.7 Inspiring from Photosynthesis and Construction of a Cycle of Molecular Code, Energy, and Mass for Sustainability

By inspiring from photosynthesis, we suggest a triangular recycling system with three major subunits of energy, mass, and code. All elements in this triangle can convert to each other via clusters of photochemical/chemical reactions (please see Figs. 4.9, 4.15 and 4.16).

Nature applies a highly efficient and smart approach for production of carefully designed masses and complex materials (with a wide range of physicochemical properties) by using solar energy and simple molecules.

In this system, molecules function as information and energy storage media, By putting different pieces of nature's puzzle together, it is possible to provide innovative solutions for big challenges for sustainability of human life and climate change.

Currently, the major focus for application of solar energy as a sustainable solution has been the production of energy in different forms of electricity, hydrogen,

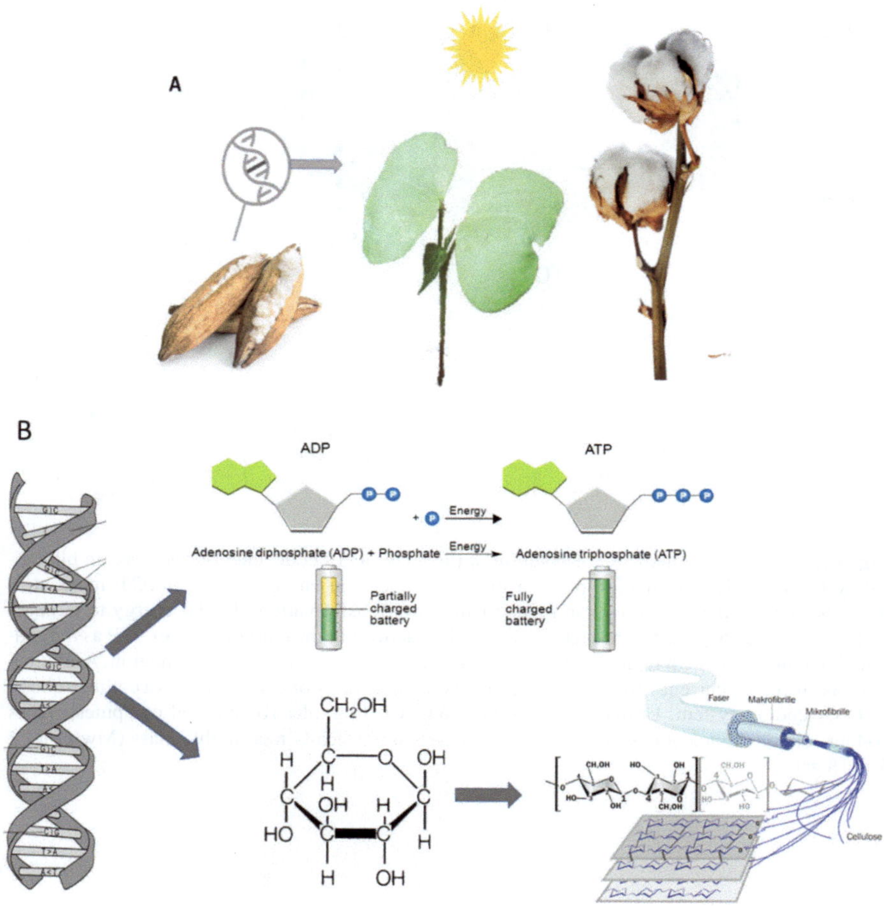

Fig. 4.14 (**a, b**) Schematic image representative of code-based energy sustainability and relativity of code, energy, and mass other than relativity of energy and mass. The whole photosynthesis process can be simulated in synthetic forms. But, in the current design of solar cells and artificial photosynthesis systems, the full potential of plants photosynthesis has not been used

and biofuels [2, 7, 10, 11, 14, 15, 25]. Less focus has been on the production of tailored designed mass from solar energy. Generation of such biologically inspired systems and materials that can mimic the properties of biomaterials via relativity in information, energy, and mass transformation will open new sustainable and clean solutions for human life.

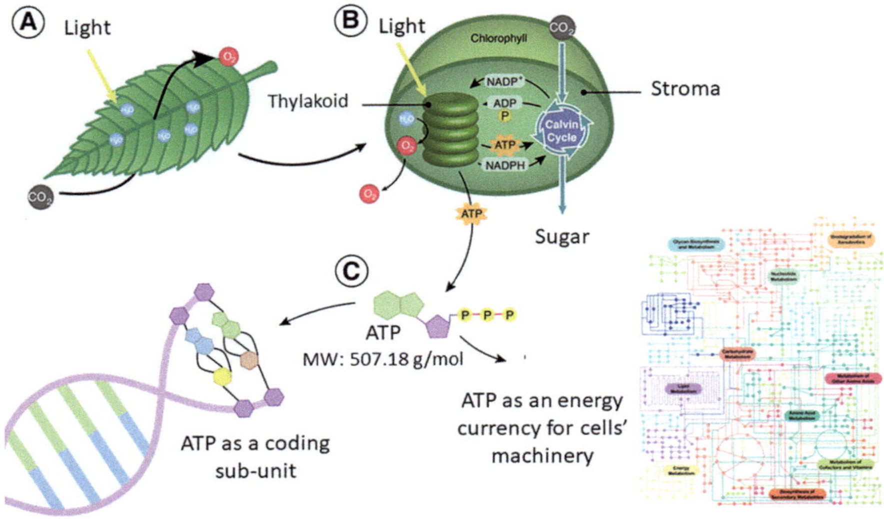

Fig. 4.15 (**a–c**) A schematic representation of the relativity of code, energy, and mass in biological systems. (**a**) Transformation of solar energy to chemical energy (stored in ATP molecules) through photosynthesis reactions in green plants. (**b**) Transformation of solar energy to biomass (synthesis of carbohydrates) through Calvin cycle reactions in the chloroplast. (**c**) ATP as a multifunctional molecule with relativity in coding capacity, energy, and mass transformation. ATP functions as an energy currency for running the metabolic machinery of cell, as well as a coding subunit in DNA coding system. Unlike binary coding units (0, 1) in electronic-based computers, DNA coding subunits (nucleotides including ATP) possess a physically measurable mass (Mw of ATP, 507.18 gr)

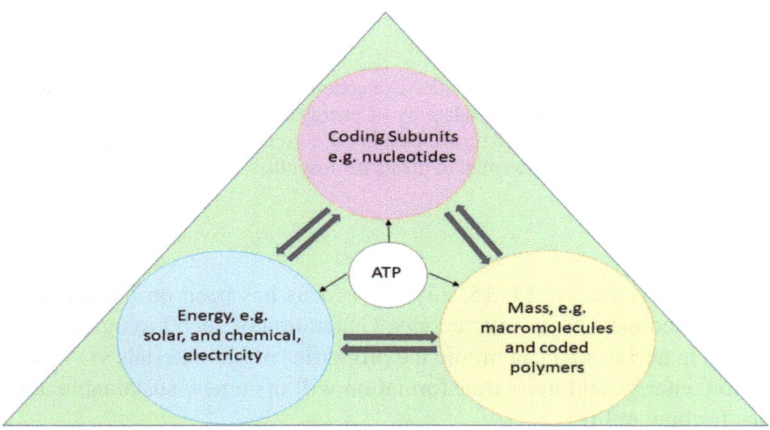

Fig. 4.16 A schematic image representative of free conversion of subunits of energy, mass, and code to each other in a living cell. ATP plays a central role in this process

References

1. Alberts B, Johnson A, Lewis J, Raff M, Roberts K, Walter P (2008) Molecular biology of the cell, 5th edn. Garland Science Taylor and Francis Group, New York
2. Armaroli N, Balzani V (2011) Towards an electricity-powered world. Energy Environ Sci 4:3193–3222
3. Barber J, Tran PD (2013) From natural to artificial photosynthesis. J R Soc Interface 10:1–16
4. Blankenship R, Chen M (2013) Special expansion and antenna reduction can enhance photosynthesis for energy production. Curr Opin Chem Biol 17(3):457–461
5. Carlowcz M, Simmon K (2012) NASA Earth Observatory (2012), seeing forests for the trees and the carbon: mapping the world's forests in tree-dimension. Accessed Jan 2012
6. Chin AW et al (2013) The role of non-equilibrium vibrational structures in electronic coherence and recoherence in pigment-protein complexes. Nat Phys 9:113–118
7. Cosnier S, Holzinger M, Le Goff A (2014) Recent advances in carbon nanotube-based enzymatic fuel cells. Front Bioeng Biotechnol 2(45):1–5
8. Dortman KE et al (2013) Photosynthetic reaction center as a quantum heat engine. PNS 110(8):2746–2751
9. Engel GS et al (2007) Blankenship & G. R. Fleming, 'Evidence for wavelike energy transfer through quantum coherence in photosynthetic systems'. Nature 446(7137):782–786.
10. Fukuzumi S (2015) Artificial photosynthetic systems for production of hydrogen. Curr Opin Chem Biol 25:18–26
11. Graetz J (2009) New approaches to hydrogen storage. Chem Soc Rev 38:73–82
12. Gruber JM et al (2018) From the isolated light-harvesting complex to the thylakoid membrane: a single molecule perspective. Nano 7(1):81–92
13. Hildner R et al (2013) Quantum coherent energy transfer over varying pathways in single light-harvesting complexes. Science 340:1448–1451
14. Huntley ME, Redalje DG (2007) CO2 mitigation and renewable oil from photosynthetic microbes: a new appraisal. Mit Adapt Strat Glob Chang 12:573–608
15. Kalyanasundaram K, Graetzeel M (2010) Artificial photosynthesis: biomimetic approaches to solar energy conversion and storage. Curr Opin Biotechnol 21(31):298–310
16. Khouly M, Mhsnawy E, Fukuzumi S (2017) Solar energy conversion: from natural to artificial photosynthesis. J Photochem Photobiol C: Photochem Rev 31:36–83
17. Kim D, Sakimoto K, Hong D, Yang P (2015) Artificial photosynthesis for sustainable fuel and chemical production. Angew Chem Int Ed 54:2–10
18. Lambert N et al (2013) Quantum biology. Nat Phys 9:10–17
19. Lodish H et al (2013) Molecular cell biology, 7th edn. W.H. Freeman, New York
20. Majumdar P, Pant D, Patra S (2017) Integrated photoelectrochemical systems: a paradigm shift in artificial photosynthesis. Trends Biochem 35(4):285–287
21. Mcconnel I, Li G, Brudvig G (2010) Energy conversion in natural and artificial photosynthesis. Chem Biol 17(5):434–447
22. NASA Earth Observatory (2012) Seeing forests for the trees and the carbon: mapping the world's forests in three-dimension. Accessed Jan 2012
23. Nelson DL, Cox MM (2017) Lehninger principles of biochemistry, 7th edn. Freeman, W. H. & Company, New York
24. Niederholtmeyer H, Wolfstadter BT, Savage DF, Silver PA, Way JC (2010) Engineering cyanobacteria to synthesize and export hydrophilic products. Appl Environ Microbiol 76:3462–3466
25. Orimo AI et al (2007) Complex hydrides for hydrogen storage. Chem Rev 207:4111–4132
26. Panitchayangkoon G et al (2011) Direct evidence of quantum transport in photosynthetic light-harvesting complexes. PNAS 27(54):20908–20912
27. Rodwell VW, Bender D, Botham K, Kennelly P, Weil PA (2015) Harpers illustrated biochemistry, 30th edn. The McGraw Hill Education, New York
28. Romero E et al (2014) Quantum coherence in photosynthesis for efficient solar- energy conversion. Nat Phys 10:676–680
29. Verseux C et al (2016) Sustainable life support on Mars-the potential roles of cyanobacteria. Int J Astrobiol 15(1):65–92

Chapter 5
The Relativity of Code, Energy, and Mass: The Molecular Mechanism of Autonomy and Self-Organization in Biological System

Abstract Biomolecules are the principal building blocks from which a cell is constructed. A large number of functional biomolecules, exciting in the cells, are in the form of polymeric macromolecules composed of sequences of various monomeric subunits. Still, biomolecules have remarkable properties that make the chemistry of living systems special compared to other non-living chemical elements in nature. A special property of biomolecules is related to the information storage and transduction via the electrochemical interactions among their atoms.

This causes the multifunctional property of materials (matter) in living cells, as units of energy, mass, and information storage/transfer at the same time. This multifunctional property of matter in living cells can be observed at different scales, starting from subatomic level (quantum) to molecular, micro, and macro levels of complexity.

In fact, the high efficiency and autonomous nature of biological processes in information operation as well as, energy and mass transfer are directly related to the multifunctional property of biomolecules (biomatter). In previous chapters, we discussed each of these features of autonomy in living systems, including self-regulation, self-organization, and self-fueling properties, separately.

In this chapter, we attempted to describe the integrity among molecular mechanism of endogenous information processing, self-fueling, and self-organization.

Biomolecules are able to shift among their functions (as units of information storage, energy, and mass) and produce freely based on the dynamic needs of a living system in interaction with its environmental factors.

In this chapter by referring to their highly flexible chemical structure, we attempted to describe the multifunctional properties of biomolecules. To this end, we discussed the integrity of metabolic pathways among three major categories of biomolecules, including nucleic acids, carbohydrates, and amino acids.

Multifunctional property of biomolecules (as units of information, energy, and mass), as well as their structural flexibility (through the integrity of metabolic pathways), also called the relativity of code, energy, and mass, describes several features of autonomy in biological systems.

By inspiring from biological systems, relativity of code, energy, and mass in biomolecules can be applied in the future as an efficient solution for designing and

T. Karimi, *Molecular Mechanisms of Autonomy in Biological Systems*,
https://doi.org/10.1007/978-3-319-91824-2_5

generation of synthetic autonomous system as well as a sustainable solution for production of smart and programable materials.

Keywords Autonomy · Programmable materials · Biomolecules · Information-enriched macromolecules · Quantum mechanics · Relativity of code · Energy and mass

5.1 Introduction

As we discussed in previous chapters, autonomy is an inherent property of living systems. Autonomy in biology means that a system is existing and functioning as an independent system.

In fact, autonomy in a living organism involves three aspects which lead the independent existence of the system, including:

1. The endogenous information processing (self-regulation) and self-recognition of environmental signals
2. Self-sustaining energy uptake from the environment (self-fueling and self-sustainability)
3. Self-regulated formation and assembly of system's units (self-organization)

In other words, life has been designed by nature for self-regulation, self-sustainability, and self-organization.

In pevious chapters we discussed each feature of autonomy in living systems, separately (including self-controlling, self-organization, and self-fueling properties). However, still, there are several features of autonomy in biological system that cannot be explained if we consider the regulatory mechanisms of information operation and energy and mass transformation in isolation.

In fact, molecular mechanisms behind self-organization, endogenous cognition, and energy management in living cells are highly integrated. Keeping the highly organized and information-enriched structure of cells is not possible without the continuous energy uptake from the environment. On the other side, autonomous energy transfer in living cells is precisely programmed through the highly interconnected genetic and metabolic pathways (including DNA, RNA, protein, hormones, etc.).

Here, in this chapter, we attempted to merge the overlaying molecular mechanisms behind endogenous cognition, self-fueling, and self-organization, to determine a new definition for autonomous behavior of living systems. This new definition can help us fill the gaps between the cognition and physical properties (or the gap between the mind and body) in biological systems.

As we discussed in Chap. 2, at the early twentieth century, Erwin Schrodringer attempted to provide a new systametic definition of *life* through basic principles of quantum physics in his famous book, titled *What Is Life?*

In this book, Schrodringer introduced the idea of "aperiodic crystal" that carries genetic information. Eventhough the existence of heritable information had been hypothesized since 1869, the chemicophysical structure of genetic material still was not known when Schrodringer predicted the existence of *aperiodic crystal* as an inheritable material in living cells. Later, in the 1950s, the discovery of double helix of DNA by Watson and Crick confirmed the theory of "aperiodic crystal" by Schrodringer on genetic material.

> *We must therefore not be discouraged by the difficulty of interpreting life by the ordinary laws of physics. For that is just what is to be expected from the knowledge we have gained of the structure of living matter. We must be prepared to find a new type of physical laws prevailing in it'.*
>
> *Erwin Schrodinger*

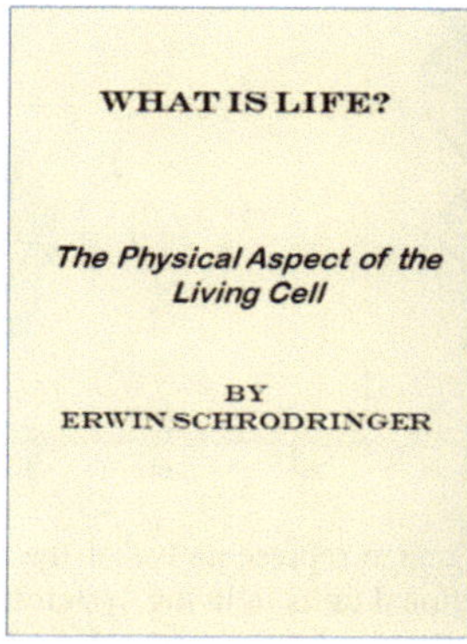

What is still less known, over half a century of work, inspired by Schrodinger's definition of life, as "ordered physics," is that: How autonomous behavior of biological systems can be explained by harmonic interactions of thousands of biomolecules (not only DNA) with each other as well as physicochemical environmental factors. How does nature apply chemistry of life as a matrix for coding and data operation? How nature integrates the body and mind in the chemistry of a living cell? (Unlike any human-made autonomous system, there is no gap between the physical mass and the programming system of a living cell.)

How can we learn from nature's design for generation of molecular-based autonomous systems?

Can we explain all properties of biological systems by our current understanding of natural sciences through physics, chemistry, and thermodynamics?

By considering the physical and chemical foundation of living systems, what is specific about the chemical and physical foundation of life? In ths chapter we attempted to connect the dots between mechanisms of information, energy and mass transfer in living cells.

According to the description of Erwin Schrodinger for life, biological systems obey basic principles of physics (including relativity of energy and mass, as well as quantum mechanics) [14].

More recent studies on the functional properties of living cells at submolecular levels also reveal that quantum mechanics is harnessed by living systems in several ways to obtain their biological advantages. For example, fundamental biological processes that involve the conversion of light energy into forms of more stable chemical energy are quantum mechanics in their nature [2, 4, 11]. These processes involve light absorption, electron excitation, transfer of electrons and photons, as well as, energy and mass transformation. Photosynthesis [2, 4, 11] and self-navigation capacity of birds in the magnetic field of the earth are examples that reveal the occurrence of quantum mechanics in biological systems [6, 12, 15].

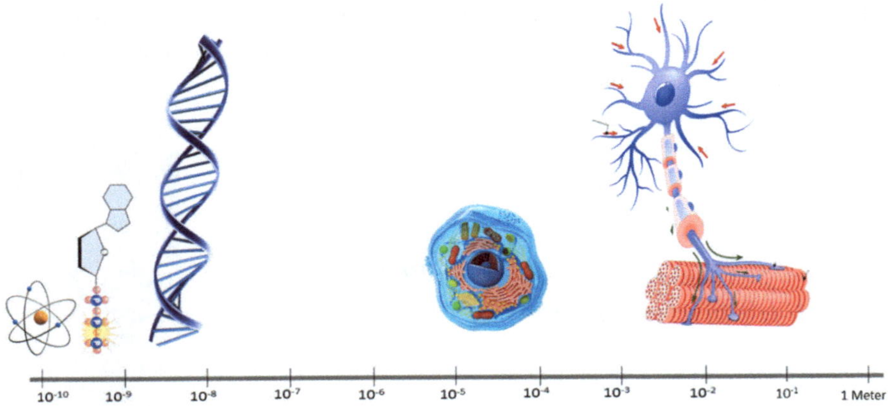

Fig. 5.1 Schematic image representative of the multitasking behavior (energy and information transfet) among functional units in living systems, starting from subatomic (quantum) level to the molecular, micro, and macro levels of complexity

We know that living systems harness several aspects of quantum mechanics (e.g., quantum coherence effect in light-harvesting complex) and formation of excited electronic states transfer of electrons and protons in chemical processes such as photosynthesis and cellular respiration. But, still, biological systems indicate remarkable features of quantum mechanics that distinguish them from inanimate elements and non-living systems. Below, we determined a list of distiniguishing properties of biomolecules:

1. The multitasking behavior of functional units of living systems, starting from subatomic (quantum) level to the molecular, micro, and macro levels of complexity. Figure 5.1 illustrates a schematic image representation of energy and information transfer by biomatter at different scales of complexity, from subatomic, atomic, molecular, micro, and macro levels of complexity.
2. Free transformation of subunits of code, energy, and mass to each other (e.g. through the integration of metabolic pathways of nucleic acids, carbohydrates and amino acids).
3. Autonomous formation of a designer mass from less organized materials (self-organization). Figure 5.2 illustrates a schematic image representation of conversion of energy of light photons to the chemical energy of biomolecules including ATP and glucose (part A, B). The process continues with the autonomous formation of a designer mass applying less organized molecules (e.g., CO_2, H_2O) and solar energy (part C).
4. Quantum behavior of biomaterials (biomolecules) at room temperature and aqueous environment (e.g., quantum tunneling behavior of enzymes and quantum coherence of molecules in light-harvesting complex of photosynthesis).
5. High flexibility in the three-dimensional structure of signaling proteins under the remote effect of physicochemical environmental factors. The dynamic folding of proteins leads their logic gate activity and signal transduction (please see Chap. 2 for detailed information).

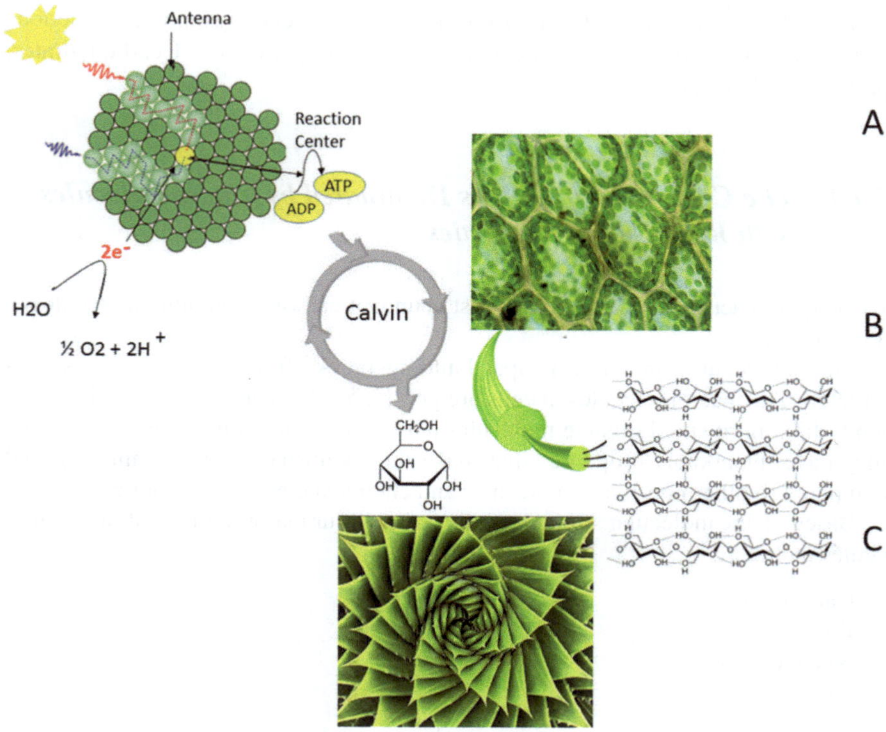

Fig. 5.2 Schematic image representative of free transfer of energy and mass in plants' cells, followed by autonomous production of designer mass. (**a**) Absorbtion and transfer of energy of photons by a cluster of antenna-harvesting complex through quantum coherence effect. Solar energy is used to split molecules of water and also to produce ATP. (**b**) In the next stage, Calvin cycle, carbon fixation occurs and glucose as a base nutrient is produced. (**c**) Through the polymerization of glucose molecules, cellulose fibers are generated. Self-organization of cellulose fibers and other biomolecules leads the autonomous formation of a designer mass

6. The existence of the internal cognition in collective behaviors among different subunits of a living system. This property can be observed in the autonomous formation of signaling networks at different levels of complexity (e.g., formation of intercellular protein signaling networks, the formation of cell-cell interaction networks, collective migration of cells and formation of dynamic patterns during embryonic morphogenesis, as well as the formation of dynamic patterns of neural networks in the brain).

To address the above mentioned challenging properties of biomaterials, in the following part, firstly we provide an overview on the multifunctional properties of three major classes of biomolecules including carbohydrates, proteins, and amino acids, which are well defined as major units of energy, mass and information transformation in living organisms. Then, we will discuss the flexibility of biomolecules to shift among different physiological functions, based on the dynamic needs of the system. Flexibility of biomolecules to shift among different physiological functions can be described, through the integrity of metabolic pathways of carbohydrates, proteins, and nucleic acids.

5.1.1 The Chemistry of Cells Is Dominated by Macromolecules with Remarkable Properties

By weight, macromolecules are the most abundant carbon-containing molecules in a living cell.

Macromolecules are the principal building blocks from which a cell is constructed. The macromolecules in cells are polymers which are constructed by covalently linking of small organic molecules (called monomers) into long chains. Yet, they have remarkable properties that make the chemistry of living things special compared to other non-living molecules and chemical elements in nature.

Based on the molecular structure, cells contain four major families of macromolecules including:

1. Carbohydrates
2. Proteins
3. Nucleic acids
4. Fats

Having looked at the ways atoms combine into small molecules and how these molecules behave in an aqueous environment, we describe the structural and functional properties of these main classes of macromolecules in living systems.

If we disregard water and inorganic ions (e.g., K^+ and Na^+), nearly all the molecules in a cell are based on carbon. Carbon is outstanding among all the elements in its ability to form large molecules. Carbon atom is small itself but has four electrons with the capacity to form four covalent bonds with other atoms. Most importantly, one carbon atom can join other carbon atoms through highly stable covalent C-C bonds to form chains and hence generate large and complex molecules with no obvious upper limit to their size. Certain combinations of atoms, such as the methyl ($-CH_3$), hydroxyl (-OH), carboxyl (-COOH), ketone (-C=O), phosphate ($-PO3^{2-}$), sulfhydryl (-SH), and amino (-NH2) groups, occur repeatedly in organic molecules. Each such chemical group has distinct chemical and physical properties that influence the behavior of the molecule in which the group occurs [1, 10].

Now let's review carbohydrates, amino acids, and nucleic acids (as the three major classes of macromolecules in biological systems) through their special structural and functional properties. Here, we emphasize on a common property of all biomolecules with regards to their multifunctional properties and high structural flexibility, as units of coding, energy, and mass in a living cell. The structural flexibility of units of in living cells (here called the relativity of code, energy, and mass) make the chemistry of cell a very efficient and programmable matrix for information, energy, and mass transfer.

5.1.2 Carbohydrates (Saccharides)

Carbohydrates (also called saccharides) are the first source of energy for living cells. Plants can synthesize carbohydrates through photosynthesis from CO_2 and water. Animals get carbohydrates from plants, or they can synthesize different carbohydrates from lipids, glycerol, and amino acids.

Carbohydrates exist in living cells in different forms of monomeric, dimeric, or polymeric (defined as monosaccharides, disaccharides, and polysaccharides) (Figs. 5.3 and 5.4).

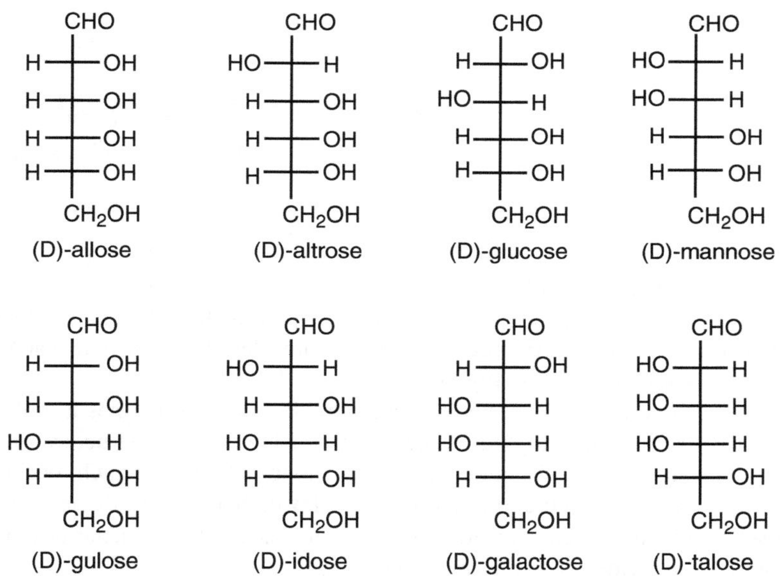

Fig. 5.3 Schematic representations of the chemical structure of common natural monosaccharides (aldoses). (Source of image: http://learnbiochemistry.blogspot.com/2014/09/hexose.html)

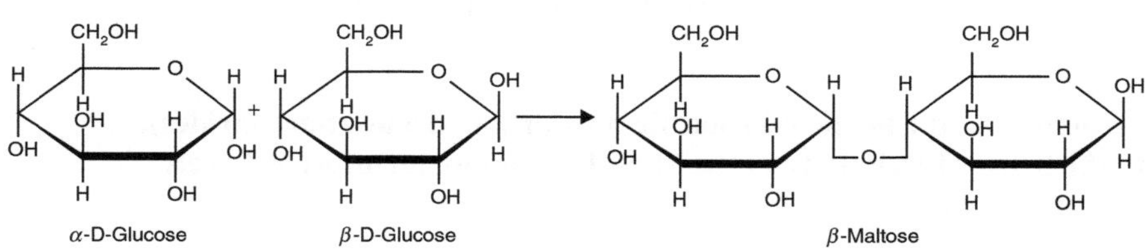

Fig. 5.4 Schematic images representation of glycoside bonds between two molecules of D-glucose and formation of maltose (disaccharide). (Source of image: http://www.chem.latech.edu/~deddy/chem121/Carbohydrates.htm)

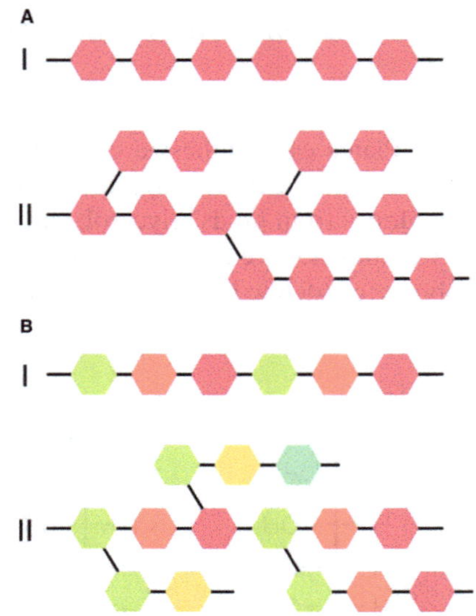

Fig. 5.5 A schematic representation of polymerization of monosaccharides by the glycoside bonds between monosaccharides [10]

The chemical structure of carbohydrates is made of aldehyde or ketone derivatives of polyhydric alcohols. Depending on whether they have ketone or aldehyde group in their chemical structure, saccharides are classified into two categories including aldoses and ketoses (Fig. 5.4).

Glucose is one of the main basic subunits of saccharides and has a molecular weight of 180.156 g/mol. Figure 5.3 demonstrates the chemical structure of main aldose monosaccharides. Polymerization of monosacharides which occurs by o-glycosidic bonds, leads the formation of complex carbohydrates. Figure 5.5 illustrates schematic image representative of different classes of polysacharides including homopolyscharides and heteropolysacharides.

Because of the polymerization reaction between monosaccharides, different variants of polysaccharides form for different biological purposes such as coding (in the structure of antigens and receptors), structure formation (such as cellulose), or energy storage (such as starch and glycogen) [13] (Figs. 5.5 and 5.6).

Structural and Chemical Properties of Carbohydrates (Saccharides), in the Sense of Relativity in the Mass Formation at Different Levels of Complexity

Using the chemical structure of glucose as an example, in the following parts, we will describe the multifunctional properties of carbohydrates in the sense of information storage capacity, as well as energy and mass transformation. To this end we will describe the biochemical reactions related to the oxidation of glucose which is

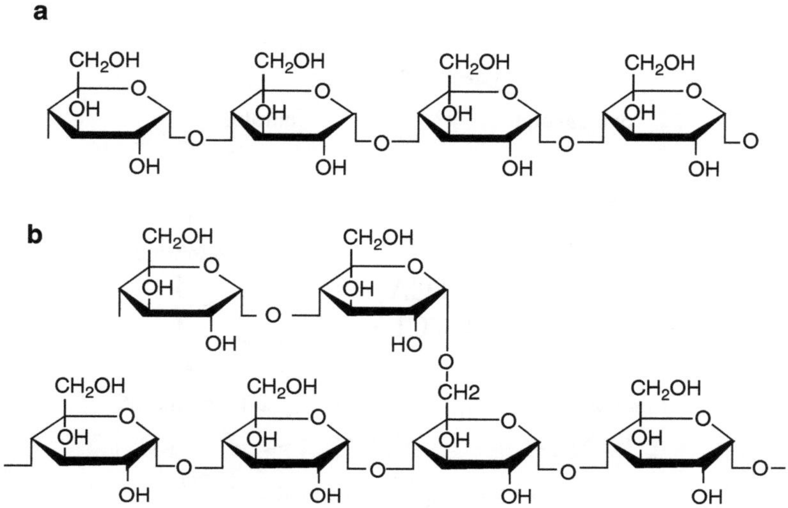

Fig. 5.6 (**a–b**) A schematic representation of polymerization of monosaccharides and formation of linear and branched polysaccharides. A schematic representative of the formation of saccharides bonds in amylose (**a**) and amylopectin (**b**). A comparison of chemical structure of amylose and amylopectin. (Source of image: http://www.rsc.org/Education/Teachers/Resources/cfb/Carbohydrates.htm#2)

considered as a main source of energy production in living cells. In addition, we will describe the coding capability of carbohydrates independently or their capability for transforming into the subunits of other coding layers such as nucleic acids and amino acids.

Energy Production Capability of Carbohydrates, in the Sense of Relativity in Energy

Glucose and other saccharides can be oxidized to produce energy. For instance, in biological systems, oxidation of each molecule of glucose results in the production of 38 molecules of ATP (which is defined as the energy currency in living systems). Oxidation of glucose in the living system occurs through sequential biochemical and enzymatic reactions, called glycolysis and the citric acid cycle [10] (Fig. 5.7).

Coding Capability of Carbohydrates, in the Sense of Relativity in Code

Glucose and other saccharide subunits such as fructose, galactose, mannose, etc. have the capability of information storage. This special layer of coding is possible via highly specific sequences of glycosides. For example, through the posttranslational modification of proteins and formation of highly specific glycoproteins which play a critical role as the major mediators of signal transduction in biological systems [7].

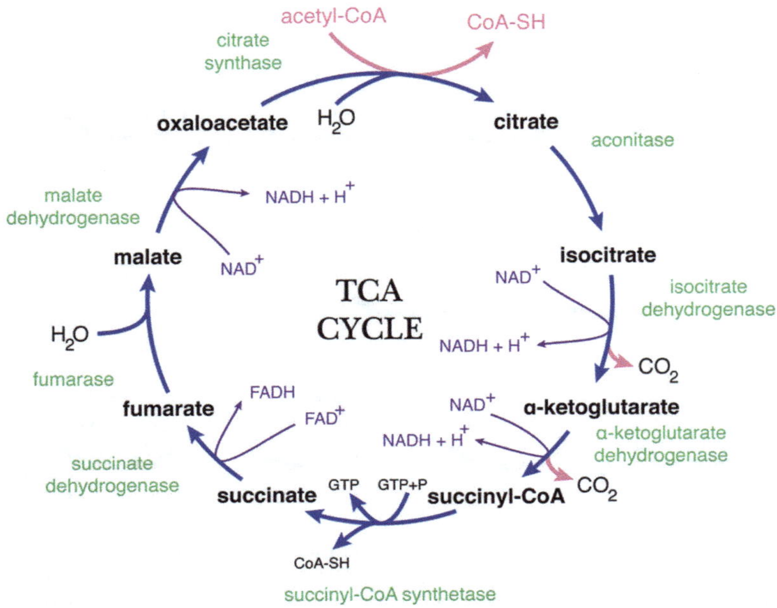

Fig. 5.7 (**a**, **b**) A schematic representative of biochemical reactions involved in the citric acid cycle, oxidation of glucose, and production of ATP. (Source of image: https://en.wikipedia.org/wiki/File:Citric_acid_cycle_with_aconitate_2.svg)

In addition, carbohydrates are structurally stable enough to be transferred to subunits of other layers of coding such as nucleotides and amino acid coding layer through the citric acid/uric acid/aspartate cycles/shunts (Fig. 5.8).

5.1.3 Proteins

Proteins are complex, organic macromolecules composed of amino acid subunits. There are 20 different forms of amino acids in the nature (Fig. 5.9). Polymerization of different sequences of amino acids by peptide bonds makes a wide variety of proteins (Fig. 5.10). The variety in chemical structure of proteins is much higher than any other organic components.

As an example, we described the chemical structure of amino acids and their polymerization capability (in the sense of mass formation at different levels of complexity) as well as biochemical reactions related to the oxidation of amino acids (in the sense of energy production capability). In addition, we described the information storage and signal transduction capacity of proteins through the sequences of amino acids and their dynamic conformational structure.

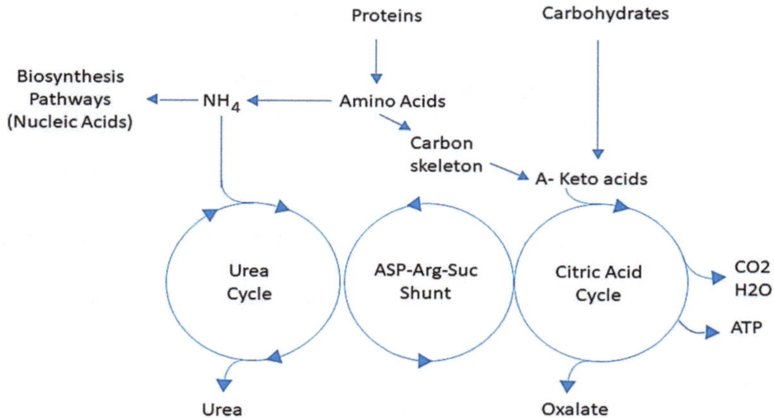

Fig. 5.8 The relativity of code, energy, and mass through the Aspartate Argininosuccinate shunt of citric acid and urea cycles. Catabolism of amino acids through the uric acid circle and its connection to the citric acid cycle (energy production cycle) through the aspartate shunt (Aspartate is a common intermediate of citric acid and urea cycles [10])

Structural and Chemical Properties of Amino Acids, in the Sense of Relativity in the Mass Formation

Figure 5.9 demonstrates the chemical and spatial structure of 20 amino acids that are defined as the subunits of proteins. Figure 5.10 demonstrates the polymerization reaction of amino acid subunits by the formation of peptide bonds [14].

Because of the polymerization reaction between amino acids, different forms of peptides and proteins form at different levels of complexity and with a wide range of functions including structure formation (such as cytoskeleton proteins), signal transduction, and coding (such as enzymes and mediators of biological signaling pathways) [10]. In addition, proteins and amino acids are flexible in the sense of energy production through specific biochemical reactions such as gluconeogenesis (e.g. energy production capacity of stored proteins in a chicken fertilized egg) (Fig. 5.11).

Energy Production Capacity of Amino Acids, in the Sense of Relativity in Energy

Amino acids can be oxidized to produce energy. For instance, in biological systems, amino acids can transform to glucose molecules through the processes and enzymatic reactions defined as gluconeogenesis (Fig. 5.11). During the sequential biochemical reactions, such as citric acid cycle, ATP is produced through the oxidation of glucose [1] (Fig. 5.11).

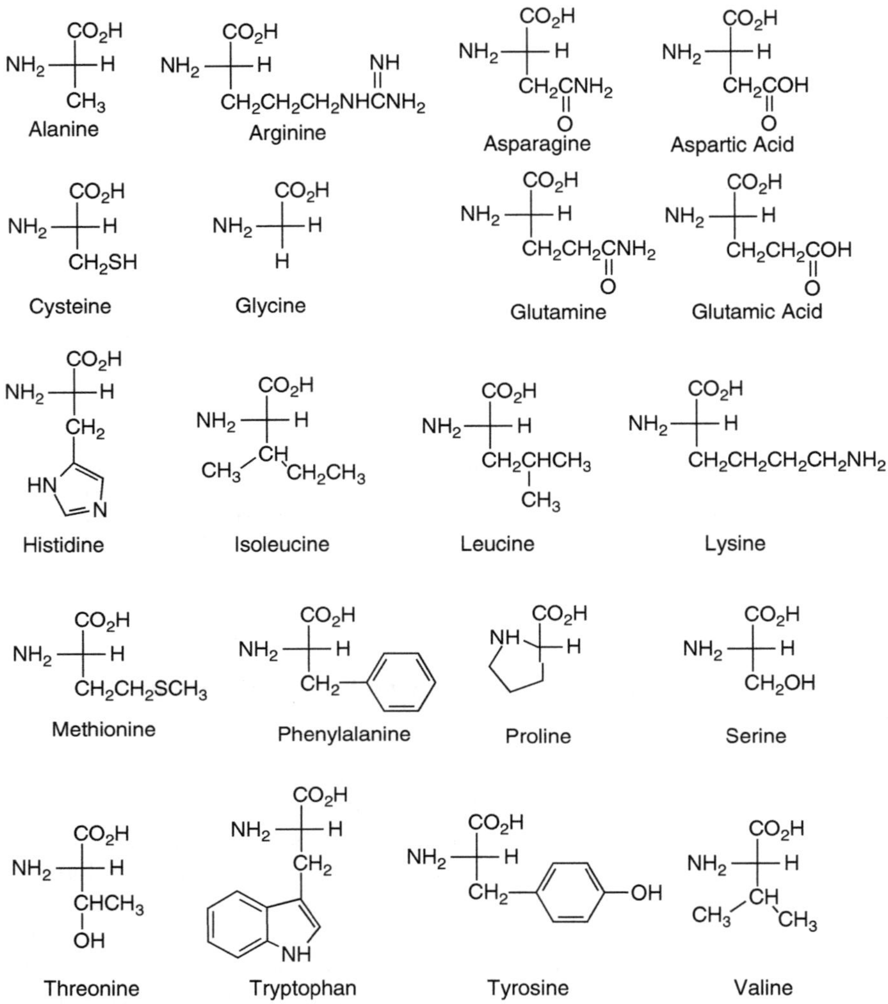

Fig. 5.9 Figure illustrates the molecular structure of 20 natural amino acids

Coding Capability of Amino Acids, in the Sense of Relativity in Code

As we discussed in Chap. 2, sequences of amino acids function as a layer of coding in biological systems. Furthermore, amino acids are structurally flexible enough to be transferred to subunits of other layers of coding such as a nucleic acid coding layer. For instance, aspartate, glutamine, and glycine can be transformed into nucleic acid polymers by providing the nitrogen source of the purine ring of nucleotides [10] (Fig. 5.12).

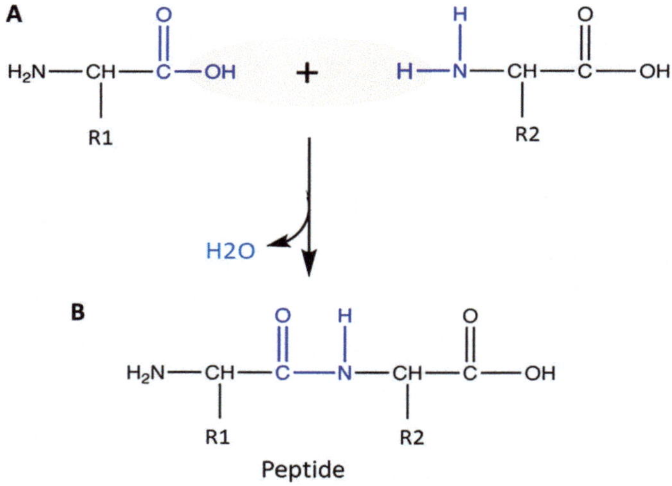

Fig. 5.10 (**a–b**) Polymerization of amino acids through the formation of a peptide bond between amino acids (**a**). Polymerization of different sequences of amino acids and formation of various peptides and proteins (**b**) [10]

5.1.4 Nucleic Acids

Structural and Chemical Properties of Nucleic Acids, in the Sense of Mass Formation

Nucleic acids are well-known biomolecules due to their main role in encoding genetic information over billions of years since the formation of a most primitive form of life. Structurally, nucleic acids are in the form of large biopolymers or small biomolecules. Nucleic acids are composed of monomeric subunits called nucleotides. Nucleotides are made of three parts, including a 5-carbon sugar, a phosphate group, and a nitrogenous base. The sugar can be ribose or deoxyribose. Depending on the type of sugar, nucleic acids are classified into two categories, including deoxyribonucleic acid (DNA) and ribonucleic acid (RNA). If ribose is applied in the composition of the molecule, the polymer is RNA. And if deoxyribose is applied in the structure of nucleotides, the polymer is DNA (Fig. 5.13 demonstrates the chemical structure of different nucleotides).

Due to polymerization reaction between nucleotides, different forms of nucleic acid polymers including DNA, mRNA, tRNA, and rRNA form with different levels of structural complexity and have different functional purposes. For instance, DNA is a highly specialized molecule for data storage, while mRNA is more specialized for data transportation, and tRNA is highly specialized for translating and converting data from one layer of coding to the next. In contrast, rRNA is specialized as a structural molecule in the formation of the ribosome.

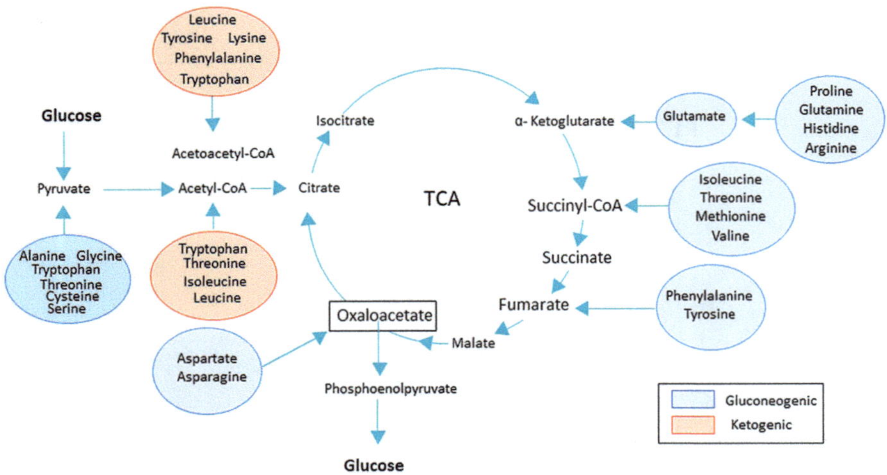

Fig. 5.11 A schematic representation of the transformation of amino acids to the glucose through the gluconeogenesis reactions. (Image adapted from Nelson and Cox [10])

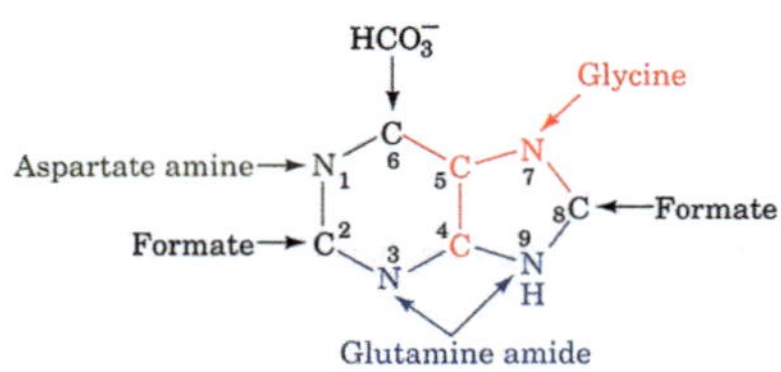

Fig. 5.12 Amino acids provide nitrogen source for purine ring in nucleic acids. In the structure of a purine, the original source of each atom is determined as follows: N1, originated from aspartate; C2, originated from formate; N3, originated from amide (N) of glutamine; C4, C5, and N7, originated from glycine; C6, originated from CO2; C8, originated from fumarate; and N9, originated from amide (N) of glutamine [10]

In addition, unlike current human-made silicon-based binary coding system, where coding subunits (e.g. bit coding units of 0 and 1 in binary coding system) cannot be measured as a specific mass with a chemical structure, nucleic acids are providing a unique coding system with a measurable mass (e.g., MW of ATP = 504.18 gr/mole).

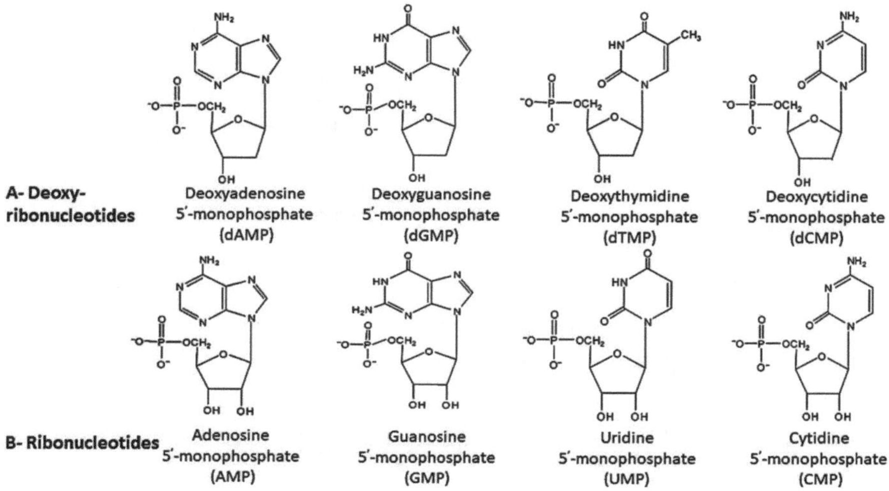

Fig. 5.13 A schematic representation of the chemical structure of nucleic acids [10]

Coding Capacity of Nucleic Acids, in the Sense of Relativity in Coding Capacity, as well as Energy and Mass Transformation

Nucleic acids are the main coding molecules in biological systems in both polymerized forms (e.g., in the forms of DNA and mRNA) and free nucleotides (such as cAMP as a major signal transducer in biological systems). Also, the double-helix structure of DNA is a molecular coding media with a specifically defined chemical structure (unlike the binary coding system which is an electronic-based coding system). For more information about the molecular- and electronic-based coding systems please see Chap. 2.

Energy Production Property of Nucleic Acids, in the Sense of Relativity in Energy Transformation Versus the Capacity of Coding and Mass Formation

Nucleic acids can be oxidized to produce energy for the metabolic tasks of cells (Fig. 5.14). ATP has been defined as the main energy currency in biological systems. ATP is produced through the cell metabolic pathways (e.g., Krebs cycle) that are oxidized of high-energy nutrients such as carbohydrates (Fig. 5.11).

In addition, degradation of amine ring of nucleotides causes the production of a-ketoglutarate and glutamate, which can be recovered to the gluconeogenesis cycle and produce glucose [10] (Fig. 5.14).

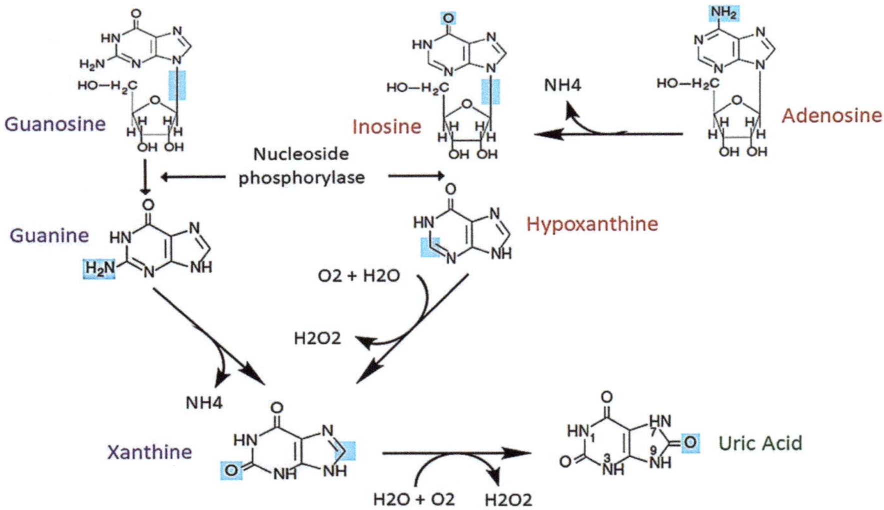

Fig. 5.14 (**a–c**) A schematic representation of degradation of nucleic acids and production of uric acid [13]

5.2 Relativity of Code, Energy, and Mass: The Main Property of Biomolecules

Here, based on the capacity of information storage in biomolecules, we attempted to define a new translation of relativity equation, as the relativity of code, energy, and mass, to describe remarkable properties of biomolecules.

The new translation of relativity equation explains that, in nature, the transformation of matter is not limited only to energy and mass but also includes molecular code as an additional factor.

The relativity of code, energy, and mass as a new concept explains the molecular regulation behind several unknown features of biological systems (such as the self-regulation and self-organization capacity). In addition, by inspiring from nature in designing biomolecules, the same basic principles can be applied to solve several challenges accress different areas of science and technology.

Here, in order to describe the concept of the relativity of code and energy as a major property of biomolecules, we referred to the multifunctional capacity of the adenosine three phosphates (ATP) molecule.

Then, based on the molecular properties of ATP (e.g., MW, 507.18 gr/mole) and the functional property of the molecule as a coding unit (in genetic coding system), we described the relativity of code, energy, and mass in biomolecules versus the relativity of energy and mass in non-living elements of nature.

As we discussed in Chap. 2, DNA is the most basic molecular coding layer in biological systems, which is made of four coding subunits of ATP, GTP, CTP, and TTP. Therefore, here, we determined one molecule of ATP as a quarter of molecular code, in DNA coding system.

Here, we used ATP because of its very well-known multifunctional capacity, as a unit of genetic code, and the main carrier of energy in living cells.

In nature, the energy of sunlight can be stored in ATP molecules by photosynthesis.

The wavelength of light used by plants in photosynthesis is about 700 nm. The energy amount stored in a single photon is given by the Planck equation at a wavelength of 700×10^{-9} m, the energy of a photon is $E = hc/\lambda$:

$$E = \mathrm{hc} / \lambda$$

$$\left[\left(6.626 \times 10^{-34}\,\mathrm{J.S}\right)\left(3.00 \times 10^{8}\,\mathrm{m/S}\right)\right] / \left(7.00 \times 10^{-7}\,\mathrm{m}\right) = 2.84 \times 10^{-19}\,\mathrm{J}$$

A mole of light is Avogadro's number (6.022×10^{23}) of photons; thus, the energy of one mole of photons at 700 nm is
(2.84×10^{-19} J/ photon) (6.022×10^{23} photons/ mole) = 17.1×10^{4} J/ mole

Therefore, production of 1 mole ATP from ADP and Pi requires 30.5 kJ/ mole energy [10]. Therefore, a mole of photons of red light has about five times the energy that is required to produce a mole of ATP from ADP and Pi. We define 1 quarter nucleotide code (for one mole of ATP molecules) as a unit of molecular code which also has 507.18 gr. equivalent level of mass. In addition ATP function as a rechargable solar battery that stores solar energy in its phosphate bonds. Here, we defined ATP as a representative of multifunctional molecules which illustrate the relativity of code, energy and mass in their physicochemical properties. This concept can be applied to other type of natural or synthetic molecules with similar capacities.

The integration of coding capacity and the thermodynamic properties (energy and mass) of biomolecules leads to the high flexibility of biomolecules for their biological function based on the system's needs. In addition, it generates a free media for the transformation of information, energy, and mass. Relativity of code, energy and mass in biomolecules can be considered as a critical element for endogenous coding which provides the internal instuction for assembling of structural units in a self-organizing system as well as, adequate flexibility for energy and mass transformation in a self-fueling system. Furthermore parallel operation capacity of coded molecules along with spontaneous electrochemical interactions between the coded molecules provides a great advantage for the operation of big data in a self-regulating molecular coding system.

Basic principles of relativity of code, energy, and mass can be applied in designing different types of bioinspired programmable materials for different biomedical and industrial applications, including:

1. Self-organizing materials and systems
2. Molecular sensors

3. Smart materials which are responsive to different types of environmental signals
4. Programmable molecules for generation of molecular computers to solve complex NP problems in logical time
5. Generation of chemical-based cognitive systems that are compatible with the human body/brain for diagnostic and therapeautic applications
6. Generation of transplantable molecular cognitive therapeutic systems, e.g. for smart targeting of cancer cells

5.2.1 Embryonic Development and Relativity of Code, Energy, and Mass

Here we considered a fertilized chicken egg as a self-organizing system with the autonomous capacity for the formation of complex geometrical structures in different tissues and organs from simpler molecular subunits (nutrients) in egg yolk.

To clarify the concept of relativity of code, energy, and mass in biological systems, first, we attempted to explain the multifunctional properties of biomolecules in a fertilized chicken egg. Then we explained the transformation capability of molecular subunits of code, energy, and mass to each other during embryonic development as a natural self-organizing system.

DNA has been defined as the main information storage and coding molecule in biological systems. Therefore, here in all calculations, we referred to DNA as the main molecular code of the system. Egg yolk is considered as the initial energy-providing mass for the system. We will follow-up the transformation trend of biomolecules to each other during the self-organization process in a growing embryo.

As a remarkable property of self-organization in a natural system, we refer to the increase in the ratio of information-enriched materials (coding molecules versus total mass) during embryonic organogenesis.

Figure 5.15 illustrates the concept of relativity of code, energy, and mass during the embryonic development of a chicken egg model system.

Physicochemical properties of self-organization can be described for the embryonic organogenesis in various vertebrate embryos. A fertilized chicken egg provides an excellent model system for studying the energy and mass transfer at different stages of this natural self-organization process. The total amount of energy and mass for an avian egg can be calculated from its dry mass at different stages of development versus the initial dry mass.

Please consider that, here, in all calculations, the physical mass of a chicken egg at each stage of development has been calculated as total dry mass to remove the effect of weight loss due to the evaporation from the egg shell [9].

In order to quantify the transformation of the subunits of energy, mass, and code to each other during the embryonic organogenesis, we calculated changes in the weight ratio of molecular code (DNA) to the total mass of the system (total dry weight of egg) at different stages of embryonic development.

First, we describe a series of stages encompassing the entire period of chicken incubation (21 days) based on chicken developmental physiology. Physiological stages of chicken embryonic development were initially explained in detail by Hamburger and Hamilton [5]. Breifly, chicken embryonic development were divided into three stage including early, middle, and late. The formation of organs and systems occur during the first two stages, followed by the organ's growth and maturation of the system at the late stage of embryonic development. Schematic Fig. 4 illustrates the timeline and onset of major functional milestones during chicken embryonic development and transformation of initial, energy-producing mass (e.g. fatty acids stored in egg yolk) to the coded materials (e.g., nucleic acids as the main biological coding materials).

Briefly, the initial stage is characterized by the development of primitive streak. The middle stage is defined primarily by the number of somites. The late developmental stage is identified by several typical morphological features. The morphological changes of limbs and visceral arches are used to determine the late stage. Toward the end of development, the chick does not undergo further morphological changes and only increase in size.

The ability of the embryo to directly respond to the environment during incubation and adjustment of its development is fundamental. Initiation of cell division machinery in the single-cell zygote is under the control of a temperature-sensitive switch. Incubation at 37 °C can stimulate the initial embryonic cell division and growth [8].

As a self-organizing system, the energy balance of organogenesis within a chicken egg is supported by oxidation of lipids, carbohydrates and proteins in egg yolk and egg white. In addition to the energy production, numerous amount of intial biomolecules in the egg transform from less-organized condition towards the synthesis of highly organized and information enriched molecules. The chemical composition of the egg is summarized as following; water comprises 74% of a chicken egg. The main chemical composition of the dry mass of egg includes 11.8% lipids, 12.8% proteins, and very small amounts of carbohydrates and minerals (less than 1%). 94% of the minerals are in the eggshell fraction, and the rest are distributed in egg white and egg yolk [3, 9]. The nucleic acid content of a fertilized egg is equivalent to the nucleic acid content of a single cell (zygote), which typically is measured in term of mass in picograms or one trillionth (10^{-12}) of a gram. During embryo development, the amount of coding subunit (e.g. DNA as genetic code) gradually increases, while the total mass of egg yolk and egg white decline over time.

Figure 5-15 also shows the ratio of molecular coding mass (DNA ratio) to the total mass increases over time, while the ratio of nutrient mass (energy subunits) to the total mass declines over time. Therefore, it can be concluded that during avian embryonic organogenesis as a natural self-organizing system, subunits of code, energy, and mass are able to transform into each other freely.

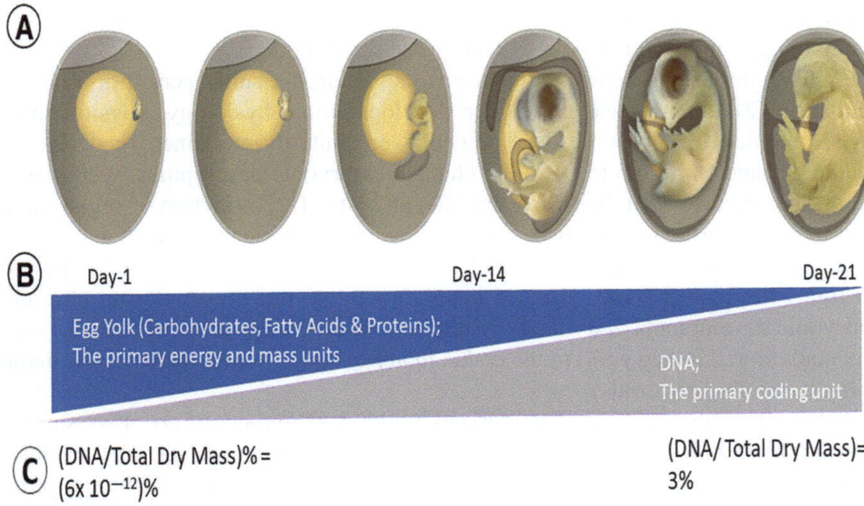

Fig. 5.15 Schematic image representative of the concept of relativity of code, energy, and mass during the self-organizing process of embryonic development by free transformation of subunits of code, energy, and mass to each other. (**a**) Schematic representative of various stages of organogenesis in a chicken embryo from day 1 (D 1) to day 21 (D 21) of development as a natural self-organizing system to the primary coding units. (**b**) Changes in the weight ratio of primary energy and mass providing materials (egg yolk and egg white) to coding units (DNA) during the organogenesis. (**c**) Transformation of subunits of energy and mass to code during the chicken embryonic organogenesis can be quantified by the measurement the weight ratio of coding molecules (DNA) to the total mass of egg at different stages of organogenesis. During embryonic development, the dry mass of coding and information enriched materials (e.g. DNA as genetic code) gradually increases while the dry mass of initial energy providing materials (e.g. carbohydrates and fatty acids in egg yolk) decline over the time. The process is regulated by integrity of metabolic pathways and the unique transformation capacity of biomaterials to each other, based on the dynamic needs of the system. Relativity of code, energy and mass is a unique property of biomaterials that provides the required structural and functional flexibly for progress of self-organization

The high flexibility of biomolecules and free transformation of molecular units of code, energy, and mass, is regulated by the cross talk between cell signaling and metabolic pathways. Synchronized harmony between cell signaling and metabolic pathways also leads the free transformation of molecular units of energy (e.g., initial fatty acids in egg yolk) to molecular coding units (e.g. DNA) in growing embryo (Fig. 5.15).

References

1. Alberts B, Johnson A, Lewis J, Raff M, Roberts K, Walter P (2008) Molecular biology of the cell, 5th edn. Garland Science Taylor and Francis Group, New York
2. Arp TB et al (2016) Natural regulation of energy flow in a green quantum photocell, Nano Letters 16(12): 7461–7466

3. Cherian G, Quezada N (2016) Egg quality, fatty acid composition and immunoglobulin Y content in egg from laying hens fed full fat camelina or flax seed. J Anim Sci Biotechnol 7:15
4. Fassioli F (2013) Photosynthetic light harvesting: excitons and coherence. J R Soc Interface 11:1–9
5. Hamburger V, Hamilton H (1992) A series of normal stages in the development of the chick embryo. Dev Dyn 195:231–272
6. Lambert N et al (2013) Quantum biology. Nat Phys 9:10–18
7. Lodish H et al (2013) Molecular cell biology, 7th edn. W.H. Freeman, New York
8. Loyau T et al (2016) Thermal manipulation of the chicken embryo triggers different gene expression in response to a later heat challenge. BMC Genomics 17(329):1–15
9. Nakano T, Ikawa NI, Ozimek L (2003) Chemical composition of chicken eggshell membranes. Poult Sci 82:510–514
10. Nelson DL, Cox MM (2017) Lehninger principles of biochemistry, 7th edn. Freeman, W. H. & Company, New York
11. Panitchayangkoon G, et al.,2011, Direct evidence of quantum transport in photosynthetic light- harvesting complex, 108 (52), 20908–20912
12. Qin S et al (2016) A magnetic protein biocompass. Nat Mater 15:217–226
13. Rodwell VW, Bender D, Botham K, Kennelly P, Weil PA (2015) Harpers illustrated biochemistry, 30th edn. The McGraw Hill Education, New York
14. Schrodinger E (1944) What is life? Cambridge university press
15. Wiltschko R, Wiltschko W (2013) The magnetic based receptors in the beak of birds and their role in avian navigation. J Comp Physiol A 199:89–98

Chapter 6
Molecular Mechanism of Autonomy and Self-Organization: An Emerging Concept for the Future of Biomedical Sciences

Abstract The whole human body can be considered as a precisely programmed system. The entire physiological functions of the human body is regulated through a hierarchical (multilayer) molecular coding system. Disorder in any part of this coding system causes misfunction in the physiological functions in the human body which also can be defined as diseases. Deep understanding of the regulatory mechanisms behind the physiological functions at different layers of these molecular coding systems can open new avenues toward the treatment of diseases with no current cure. Here, we attempt to classify different diseases based on the etiology of diseases at the molecular coding level. We provide examples of diseases with no current effective cure applying conventional therapeutic approaches.

Keywords Protein misfolding diseases · Cancer · Aging · Stem cell-based regenerative medicine · Tissue engineering · Organoid technology

6.1 Any Disease Is a Result of a Disorder in Coding Layers of Organisms

The human body can be considered as a precisely programmed system, comprised of large number of information-enriched materials. Figure 6.1a illustrates a schematic representation of the connections between different parts (organs) of the human body as a precisely programmed system. Figure 6.1b illustrates the machinery behavior of an exemplary tissue (renal tissue) which is programmed through a molecular coding system (genetic circuit in DNA). Figure 6.1c reveals the molecular regulation of a cell through genetic circuits and protein signaling molecules. Figure 6.2a–c illustrates the multilayer coding system which regulates the physiological functions in a biological system (through DNA genetic circuits, amino acids, and protein-protein interaction networks). Figure 6.2c illustrates a systemic map of metabolic networks regulating the energy balance of a cell.

The entire physiological function of the human body is regulated through a hierarchical (multilayer) molecular coding system (please see Chap. 2). Disorder in any part of this coding system results in dysfunction of the physiological functions in

T. Karimi, *Molecular Mechanisms of Autonomy in Biological Systems*,
https://doi.org/10.1007/978-3-319-91824-2_6

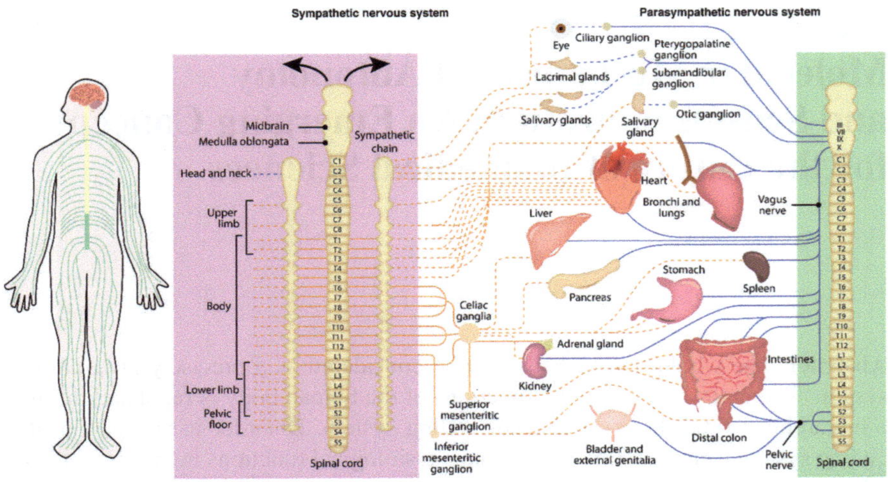

Fig. 6.1 Schematic image illustrating the whole human body function as a precisely programmed system. Similar harmony also exists at micro and nano levels among different tissues, cells, molecules, and atoms. The entire body of an organism is wired by molecules through a precisely programed chemical algorithm which has the unique capacity of signal transduction through multiple layers of complexity

the human body which also can be defined as diseases. Deep understanding of the regulatory mechanisms behind the physiological functions at different layers of these molecular coding systems can open new avenues toward the treatment of diseases that currently have no cure with conventional therapeutic approaches. Here, we attempt to classify different diseases based on the etiology of diseases at the molecular coding level.

Classification of diseases based on their coding etiology (a disorder in different layers of coding):

1. Disorders at DNA coding layer

   Examples:

   - Genetic diseases
   - Cancer
   - Viral diseases where viruses integrate DNA into their target cells such as HIV

2. Disorders in the mRNA expression

   Examples:

   - Aging-related diseases due to decline in the expression of stemness and self-renewal genes in adult stem cells

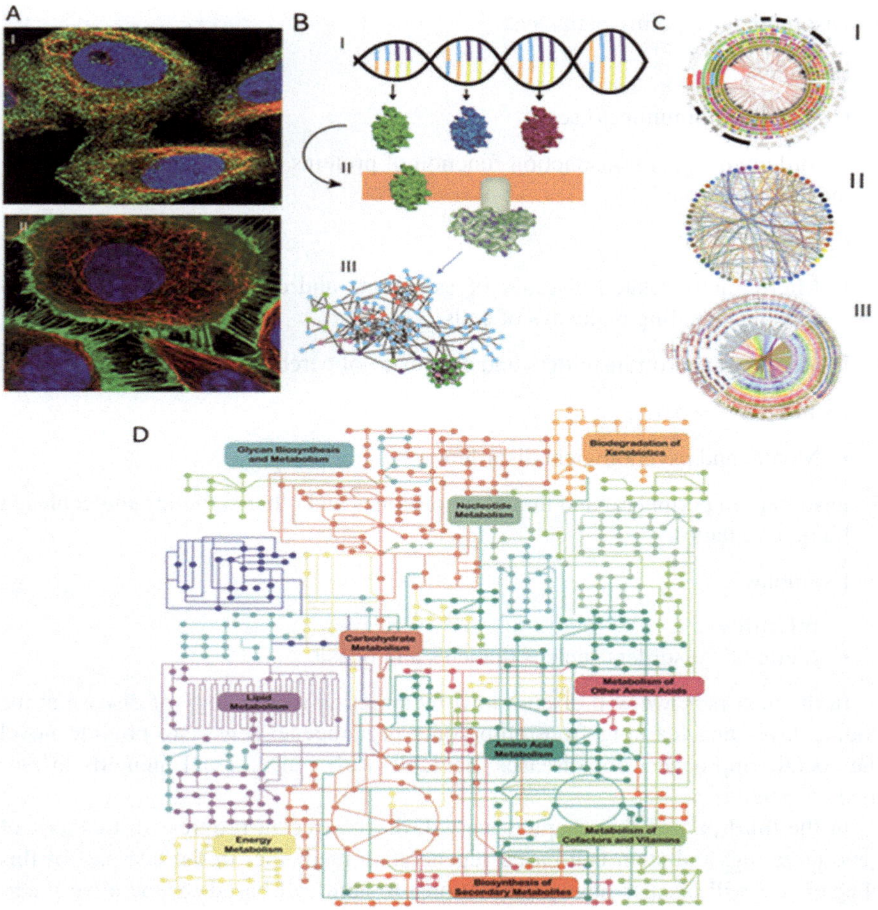

Fig. 6.2 (**a–d**) A schematic representation of the multilayer coding system in a cell through DNA genetic circuits, amino acids, and protein-protein interaction networks (**a–c**). Schematic image representative of the map of metabolic networks in a cell. (From http://manet.illinois.edu/pathways.php)

3. Disorders at protein-coding layer

   Examples:

   - Protein misfolding diseases such as multiple sclerosis (MS)
   - Age-related degenerative diseases due to protein misfolding and dysfunction of signaling and structural proteins such as Alzheimer's disease

4. Disorders in signaling pathways

   Examples:

   - Aging, autoimmune disease

5. Disorders in signal transduction function of proteins due to changes in environmental conditions

   Examples:

   - Microgravity-related diseases of astronauts and disorders in the mechanosensing signaling pathways of cells

6. Disorders in neurotransmitters and chemistry of cerebrospinal fluids (CSF)

   Examples:

   - Mental and psychological disorders

7. Disorders in peripheral and humeral hormones and other signaling molecules in blood circulation

   Examples:

   - Infertility
   - Metabolic disorders such as diabetes and obesity

In the next part, we will discuss how by targeting the etiology of disease at the coding layer and learning from embryonic organogenesis we can provide novel therapeutic approaches for diseases incurable with conventional methods in biomedical science.

In the third part of this chapter, we will discuss recent progress in methods of gene targeting, and stem cell-based regenerative medicine. In the last part of this chapter, we will discuss how combinations of stem cell based regenerative medicine, gene targeting and basic principles of the relativity of code, energy, and mass can be applied in inventions of novel therapeutic approaches by correction of coding disorders related to different diseases.

6.2 Molecular Mechanism of Organogenesis: An Emerging Therapeutic Approach for Targeting Degenerative Diseases, Aging and Cancer

As we discussed in Chap. 3, organogenesis is an autonomous process, through which all different parts of an organism form at the early stage of embryonic development. Organogenesis starts with a uniform population of embryonic stem cells (ESCs) from the inner cell mass at blastula stage. Subsequently, stem cells differentiate into all different cell types and tissues in an organism. ESCs are uniquely capable of proliferating indefinitely and differentiating into all cell types [55].

Recent progress in gene targeting, molecular, and developmental biology approaches has been applied extensively to elucidate the mechanisms of embryonic organogenesis [16, 27]. While embryonic stem cells express unlimited self-renewal capacity; after birth, all tissues and cells in a multicellular organism undergo aging due to decline in their self-renewal and regenerative capacity. Aging phenotypes have been described for stem cells of multiple tissues, including those of the hematopoietic system, intestine, muscle, brain, skin, and germline [43].

Another class of age-associated diseases are determined as degenerative diseases which are characterized by progressive degenrative changes in tissues. Alzheimer's disease is a condition determined by the progressive neural death. Age-related macular degeneration, a neurodegenerative disease, is caused by the gradual degeneration of cells in the macula of the retina. Age-related macular degeneration is the leading cause of vision loss in adults over age 55. Osteoarthritis and osteoporosis, also are other examples of age-associated diseases which are marked by the degeneration of the cartilage and bone, respectively, and cause the majority of the knee, joint, hip, and spine injuries in old patients. Degenerative diseases are defined by defects in the conformational structure of proteins or protein misfolding disorders [43, 45, 50]. Disorders in protein folding result in changes in the stability of proteins, which lead to aggregation and precipitation. Aggregation of proteins causes toxicity in the affected cells and tissues or disorders in signal transduction processes among cells. Age-associated diseases may also arise from cell dysfunction. Such conditions may include cancer and metabolic disorders such as diabetes. In the case of "adult-onset" type 2 diabetes, pancreatic islet β-cell function can be impaired such that there is insufficient insulin production or cells become resistant to insulin due to the misfunction of insulin receptors and glucose transporters [36, 43, 54].

Accumulation of age-related damages in the cell including oxidative stress can lead to DNA damage and carcinogenesis. The risk of cancer increases significantly with age [36, 55]. Regardless of their type, cancer cells share many common properties with embryonic stem cells, including fast proliferation and similar metabolic requirements. At early stages of embryonic development, stem cells gradually lose their high proliferation capacity through cell lineage specification and differentiation. After birth, adult stem cells present in different tissues are multipotent (with limited proliferation and differentiation capacity versus embryonic pluripotent stem cells with unlimited proliferation potential). In vivo studies reveal that pluripotent ESCs have inherent tumorigenic potential after transplantation into immunodeficient mice. Reprogramming of somatic cells into pluripotency by oncogenes like Myc and Klf4 demonstrates the correlations between pluripotency and tumorigenicity. Furthermore, results of cancer studies reveal that tumor tissues contain a special population of cells with self-renewal and partial differentiation capacities. Because these cells share several properties with the adult tissue stem cells from their original tissues, they are also called cancer stem cells (CSCs) [21]. In addition, the reprogramming of somatic cells and metastasis of cancer stem cells are both dependent on transitions between epithelial and mesenchymal states (EMT/MET) [4, 21].

The CSC model ultimately links cancer with stem cell biology and provides a common signature between various sources of stem cells, regardless of their developmental origin in normal or pathological states. Analysis of common regulatory mechanisms of embryonic and CSCs including gene regulatory and protein signaling pathways and epigenetic complexes can provide new solutions for regulating the destination of cancer stem cells toward the differentiation and apoptosis pathways. Additionally, knowledge about their stem cell properties is valuable for the eradication of metastatic CSCs that become resistant to radiation and chemotherapy after epithelial-mesenchymal transition [18].

Here in this chapter, we will discuss how deep understanding of molecular regulation of embryonic organogenesis (which codes the initial formation of all tissues and organs in an organism) can be applied for corrections of miscoding associated with different diseases.

We will also discuss recent advances in methods of stem cell-based regenerative medicine, and gene targeting, in biomedical sciences.

Finally, we will discuss future perspectives of biomedical sciences through the combination of stem cell-based regenerative medicine and precise genetic programming of cells (e.g., applying CRISPR-CAS9 technology).

Smart targeting of diseases at the coding level (e.g., targeting of disorders at the DNA and RNA and protein-coding levels) can open a lot of new avenues toward the treatment of diseases with no current effective cure.

6.3 Molecular Mechanism of Organogenesis and Stem Cell-Based Regenerative Medicine

During normal embryonic organogenesis, large numbers of molecular and physical signals are orchestrated in space and time to guide embryonic stem cells toward different destinations (e.g., differentiation toward different cell types and tissues). Obtaining these signals is critical in decision-making and fate determination of stem cells (Fig. 6.3) [30, 31, 37, 55].

Even though organogenesis is an internally programmed process, the system is highly connected to its microenvironment and is able to make autonomous decisions based on dynamic environmental conditions. As we discussed in Chap. 2, combinations of both conserved and dynamic coding mechanisms in stem cells lead to the decision-making capacity of stem cells by dynamic cross talk between gene regulatory networks and protein signaling networks. While gene regulatory networks provide an internal coding system for the cell, protein signaling networks are able to respond to the dynamic environmental signals. In fact, stem cells are able to make autonomous decisions by solving thousands of dynamic NP problems per

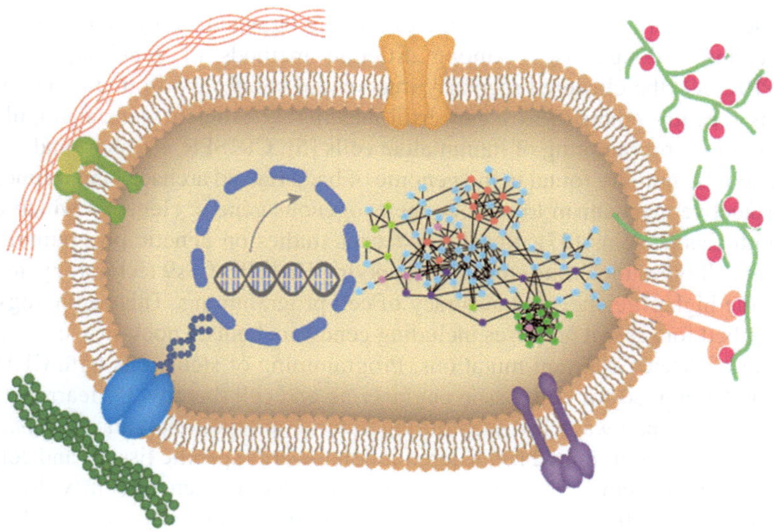

Fig. 6.3 A schematic representation of cross communication of a living cell with its environmental factors through membrane proteins and receptors. Membrane proteins also function as signal transducers and logic gates by their dynamic conformational structure. Membrane proteins transfer different physicochemical signals to the cell by changes in their conformational structure under the effect of environmental signals

second in a network as large as the entire of their genome and by considering dynamic environmental factors (Please see Chap. 2).

Stem cells retain their special qualities (e.g., stemness or differentiation potential toward various destinations) when they obtain the relevant signals from their natural microenvironment, termed "stem cell niche" [15]. In the dynamic niches of stem cells during organogenesis, stem cells integrate a complex array of molecular signals from their environment with the induced intrinsic regulatory networks to control their functions. Therefore, when we isolate stem cells from their natural niche, it is necessary to compensate the missing signals to guide stem cells toward a specific destination (e.g., differentiation into specific cell lineage).

To this end, one approach is providing a synthetic media that mimics the natural microenvironment of stem cells by adding appropriate signaling factors into this media. Tissue engineering is an interdisciplinary field that employs principles of natural tissue homeostasis in living organisms in combination with the methods of engineering and material sciences toward the generation of artificial tissue constructs (Fig. 6.3). Ultimately, the main goal of tissue engineering is to restore or improve the function of damaged or failed tissue [24, 29]. Another tissue engineering approach is internal programming of stem cells applying methods of gene targeting [55]. Targeted editing of endogenous genes, also called gene targeting, has been widely applied to program different cell types and for both clinical and basic research applications [5]. Briefly, genome-targeting involves the insertion, deletion, or replacement of DNA in the genome of a living organism using "molecular scissors" called endonucleases. Among different methods of gene targeting, the CRISPR/Cas9 (the clustered regularly interspaced short palindromic repeats) system has recently emerged as a powerful and highly efficient gene manipulation technique for programming of mammalian cells [5]. CRISPR is a short and repeating nucleotide initially found in the genome of bacteria and archaea, and it functions as a protective mechanism to remove the exogenous genetic elements which combine with Cas proteins [57]. Results of recent studies on genetic programming of different cell types including stem cells applying CRISPR/Cas9 technology demonstrated the high efficiency and accuracy of cell programming. This technology has been applied for several purposes including gene knockouts, knock-ins, and specific targeting of disease-relevant mutations. Programming of stem cells with CRISPR/Cas9 can be applied for both tissue engineering and cell therapy. By learning from gene regulatory pathways involved during natural embryonic organogenesis, stem cells have been programmed for differentiation towards specific tissues and cell lineages. Original stem cells can be easily remolded to generate new function. Furthermore, genetically engineered human immune cells and stem cells have already been applied extensively in anti-tumor research, such as expression of anti-tumor proteins and induction of apoptosis in cancer cells. Chimeric antigen receptor (CAR) T cells are considered as a long-term therapeutic approach for refractory blood cancer [5, 57].

Despite significant technical progress in the field of tissue engineering during the past decade, there are still several challenging problems to be solved. One of the key challenges is the ability to generate multicellular complex tissue structures with relevant clinical size and thickness. Another challenge is achieving the desired tissue vascular architecture [14, 17, 29]. Numerous approaches have been applied for generation of multicellular tissue structures with their physiological patterns, for example, by designing biofunctionalized scaffolds or directed assembly of organized tissue building blocks [6, 22, 24, 25]. More recently, bioprinting has been the major focus of in vitro construction of complex tissue structures [23, 32, 33]. While much progress has been made over the past decade for patterning of cells in multicellular tissue structures, embryonic organogenesis is programmed naturally to be a self-organizing process. The self-organizing capacity of stem cells through embryonic organogenesis is increasingly drawing the attention of researchers in regenerative medicine.

More recently structures resembling whole organs (organoids) have been generated from stem cells. Organoids are derived from pluripotent stem cells (PSCs) or adult stem cells (ASCs) and organ progenitors (Fig. 6.4) [27].

Generation of all different tissues and parts of an organism from a single cell (zygote) is an excellent example of self-organization and autonomy in biological systems. Embryonic organogenesis is an internally programmed process starting from a homogeneous population of stem cells that progress autonomously in a stepwise manner to differentiate a wide range of cell types in an organism [27]. Subsequent self-organization of generated cells leads to the formation of highly specific geometrical structures in different tissues and organs [55]. Organogenesis is

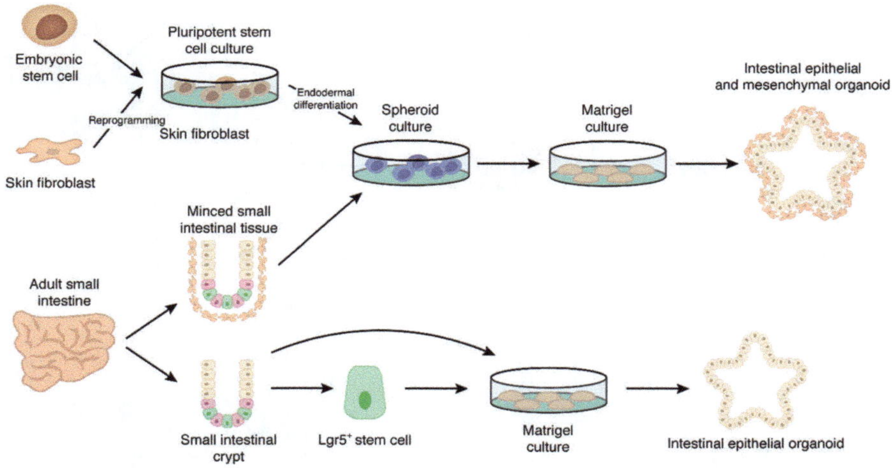

Fig. 6.4 A schematic image representative of in vitro generation of organoid tissues, applying pluripotent stem cells (ESCs and iPS cells) versus adult stem cells. (Image adapted from [26])

an internally programmed process (please see conserved coding in Chap. 2), yet, it has been equipped with another layer of dynamic coding which enables the system to make real-time decisions based on dynamic environmental signals (please see dynamic coding in Chap. 2).

Combinations of both conservative and dynamic coding systems in stem cells, designed through multiple layers of coding languages, lead to endogenous cognition and the decision-making capacity of stem cells. Therefore, during organogenesis, stem cells are able to recognize various physicochemical signals from their microenvironment and subsequently make a wide range of autonomous decisions toward different destinations based on their environmental signals.

Organoid technology is developed upon an extensive foundation of classical developmental biology. A classical approach for studying organogenesis has been to dissociate and reaggregate tissues to follow up the relative morphogenetic migration of cells in vitro (Fig. 6.5a). This approach has been applied in many classic studies with embryonic chick tissues [35, 51]. Remarkably, results of these studies reveal the capacity of dissociated cells to reorganize and segregate in a process which is called "cell sort out" to form structures with the similar morphology of the initial tissue [15, 52]. The basic principle of this type of self-assembly of cells which also is called Steinberg's differential adhesion hypothesis came from attraction and connection of cells with similar adhesive properties through the interactions between adhesive membrane proteins (Fig. 6.5a). This leads to the formation of the most thermodynamically stable structural pattern among surface adhesion proteins [7, 11, 13, 44].

The second mechanism involved in self-organizing pattern formation during embryonic organogenesis is called spatially restricted progenitor fate decisions [27] (Fig. 6.5b).

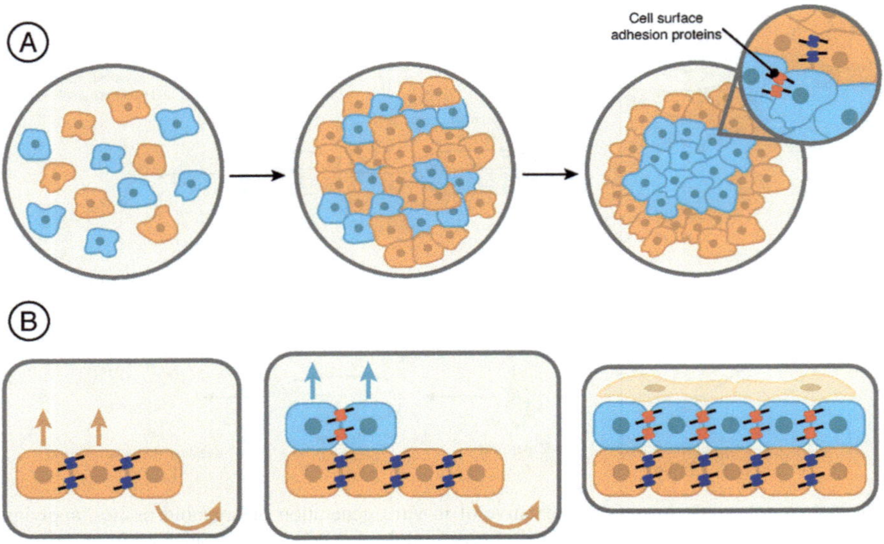

Fig. 6.5 (**a, b**) Schematic image representative of principles of self-organization during organoid formation. (**a**) Cell sort out mechanism. Different cell types sort themselves because of different adhesion properties through the interaction between adhesion molecules of the cell membrane. (**b**) Spatially restricted cell fate decision forms due to a balanced interplay between symmetric and asymmetric cell divisions and consequently migration of daughter cells toward a defined location in the tissue. (Image adapted from [27])

An excellent example of fate-decisions in stem cells under the effect of spatio-temporal environmental factors can be observed during the formation of stratified patterns in the vertebrate retina. The process starts with a neuroepithelial cell which consequently differentiates into a complex cell lineage and finally forms the multilayer (stratified) pattern of cells in the retina. The stratified pattern of retina forms due to a balanced interplay between symmetric and asymmetric cell divisions under the effect of signals from neighboring cells and extracellular matrix. Subsequent individual and collective migration of daughter cells generates tissue specific patterns [2, 56]. Reaggregation of dissociated retina cells is also a possibility in vitro condition. However, results of studies on chick embryos reveal that stratified organization of cells can be formed only when retina precursor cells are taken from an embryo before day 6 of development [40]. Remarkably, these findings indicate that retinal layering depends not only on cell sorting out but also the precise execution of fate decision by progenitor cells [27]. Therefore, it can be concluded that the combination of both cell sort out and fate decision are regulating the process of pattern formation and self-organization during embryonic organogenesis.

In particular, the self-organization property of embryonic organogenesis provides several advantages for organoid technology compared to conventional methods of tissue engineering. The autonomous capacity of specification and formation of multicellular tissues from the initially uniform population of stem cells and autonomous morphogenesis (pattern formation) are major advantages of organoid technology compared to the conventional methods of tissue engineering. In traditional

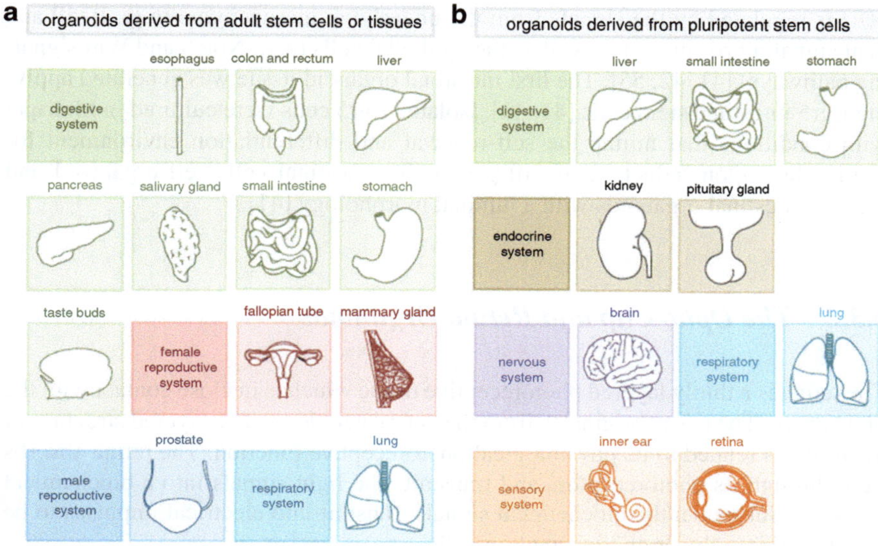

Fig. 6.6 Organoid cultures have been established for different organs and tissues applying adult and pluripotent stem cells (For detailed information, please see [26, 55]). (Image courtesy of [26])

3-D tissue engineering methods, stem cells are embedded in a natural or synthetic hydrogel. Exposing stem cells to specific culture media induces differentiation toward the desired cell lineage. However, hydrogels lack the required dynamic properties of natural ECM for exposure of signaling molecules to complete the tissue differentiation and maturation processes. In addition, these tissue constructs are far from the structural complexity of natural tissues because of the limitation in the number of cell types and methods of patterning in conventional methods of tissue engineering.

Results of recent attempts on organoid technology toward in vitro generation of tissue structures with the similar structural and functional properties of normal tissues is promising. In the following section, we will refer to some examples from the recent advances in organoid technology toward the in vitro construction of different tissues and organs including the intestine, retina, liver, kidney, neural tissues, etc. Figure 6.6 illustrates a schematic of various organoid tissues derived from adult and pluripotent stem cells (also, please see [55]).

6.3.1 Intestine Organoids

The intestine is a continuously regenerating tissue due to the existence of Lgr5 (leucine-rich G protein-coupled receptor-5)-positive intestinal stem cells [3]. The balance between the self-renewal and differentiation potential of intestinal stem cells is regulated by the signals from the underlying mesenchyme (e.g., BMP and Wnt signaling pathways) as well as the epithelial cells (e.g., Notch and Wnt signaling pathways) [41, 42, 55]. The first intestinal organoid tissue was generated applying Lgr5 intestinal stem cells [41, 42]. Isolated Lgr5 cells were cultured in Matrigel with conditions that mimic the self-renewal and differentiation environment for stem cells. Upon induction of differentiation, intestinal cells self-organized and formed intestinal organoids with a luminal morphology [41].

6.3.2 *The Optic Cup and Retina Organoids*

The retina is a thinly layered photoreceptive tissue which is in close contact with the optic nerve. The tissue originates from the neural ectoderm. The layered structure of the retina is related to its physiological photoreceptive function. The retina absorbs light through its photoreceptors and transmits the light signals into a biochemical message. Subsequently, biochemical signals transmit into electrical impulses to be transferred into the brain via ganglion cells.

In the early stages of eye development, the proliferation of cells in the germinal layer and the subsequently directed migration of the resulting cells formed the stratified structure of the retina. Results of studies on the developmental biology of the retina show that a single neuroblast precursor cell from the retinal germinal layer can give rise to at least three types of neurons in different layers of the retina [15].

In one attempt, a retina organoid was generated by recapitulating the molecular regulation of eye development in vitro. Results of this study indicated that mouse pluripotent embryonic stem cells (ESCs) can self-organize into a bilayered structure which mimics the histological properties of optic cup [8, 9]. In this study, aggregates of mouse ESCs were cultured in Matrigel containing medium with minimal growth factors. After 6 days, regions of the spherical neuroepithelium started to form lumenized structures. Results of gene expression in this study indicated the differentiating retina tissue started to express the retinal marker Rx while evaginating to form spherical buds. The distal portion of these buds subsequently underwent invagination, finally giving rise to a two-layered structure that approximates the architecture of the optic cup in vivo [16].

Results of basic studies on developmental biology of the eye revealed that several signaling pathways are involved in development of retina tissue including wingless pathway (Wnts), bone morphogenetic proteins (BMPs), fibroblast growth factor (FGF), and sonic hedgehog (Shh) [1, 10, 12, 16].

6.3.3 *Liver Organoid*

The liver originates from the endoderm. The anterior-posterior patterning of endoderm gives rise to the epithelium of the gut, midgut, and hindgut. Among them, ventral foregut contains common precursors of the liver and pancreas. The progenitors proliferate extensively, and then later at day 8.5, they differentiate into the liver or pancreas tissue based on environmental signals [38]. The initial lineage specification toward the liver and pancreas relies on gradients of FGF and BMP signals from mesoderm [27, 39].

The budding of the liver as an organ begins at E 9.5 in the mouse. Committed hepatic progenitor cells are collectively referred to as hepatoblasts, which then later give rise into two main types of liver cells including hepatocytes and cholangiocytes [19, 34, 46].

6.3.4 *Kidney Organoid*

The kidney originates from intermediate mesoderm, a subdivision of mesodermal identity that develops from the primitive streak [20, 28]. The intermediate mesoderm is specified through the action of Fgf and Wnt signaling. This tissue then specifies into two closely interacting subdomains, including the ureteric bud and the metanephric mesenchyme. Reciprocal signal transduction between ureteric bud and metanephric mesoderm promotes the outgrowth and branching of different parts of renal tubules. Specifically, ureteric bud specification can be induced by exposure of human-induced PSCs in 2D to Bmp4 and Fgf2 [53], followed by subsequent exposure of retinoic acid, Bmp2, and activin A to the cells. Generation of the second major renal precursor tissue, the metanephric mesenchyme, can be initiated by exposure of human PSCs [47] to activin A followed by Bmp4 and the Wnt agonist CHIR99021. Subsequent application of retinoic acid and Fgf9 leads to the specification of cells to the metanephric mesenchyme identity. Co-culture of ureteric bud and metanephric mesenchyme with the spinal cord tissue, which is a nephric inducer, leads the formation of well-organized nephric tubules and even nascent glomeruli (Fig. 6.7) [48, 49].

6.3.5 *Conclusion and Future Perspective*

The human body can be defined as a highly cognitive and decision-making system that is precisely programmed through multiple layers of molecular coding and is able to understand and respond to different environmental signals. At early stages of embryonic development, all different tissues and organs are forming autonomously from an initially homogenous population of pluripotent stem cells. This process which is called organogenesis is precisely programmed through multiple layers of coding.

Therefore, any disorder among different coding layers of this system causes a malfunction in operations of different tissues and organs which subsequently would be observed as symptoms of different diseases. Deep understanding of the molecular programs that regulate the formation and homeostasis of different tissues and

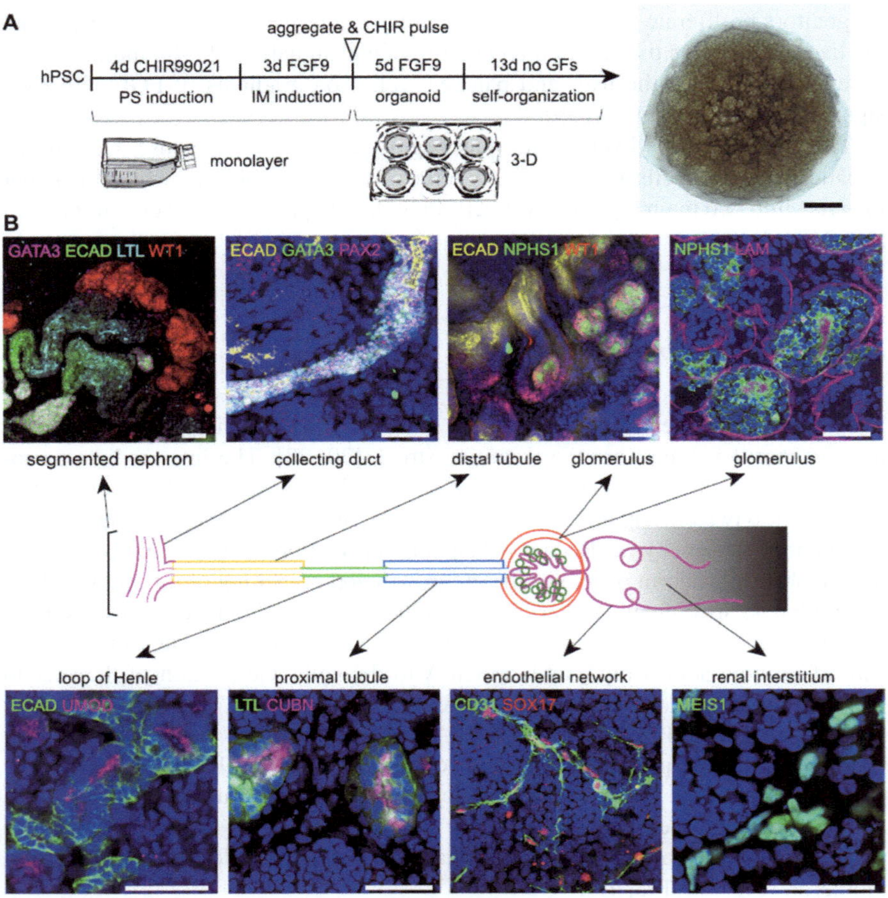

Fig. 6.7 Generation of kidney organoids from human pluripotent stem cells. A schematic image representative of a protocol for the directed differentiation of human PSCs into kidney organoids. B IHC staining shows the existence of different types of renal cells in the kidney organoids, including collecting ducts (ECAD, PAX2, GATA3), distal tubules (ECAD-positive, GATA3, and ITL-negative), loops of Henle (UMDD), proximal tubules (LTL CUBN), and glomeruli (NPHS1, WT1), with basement membrane (LAM), endothelial network (CD31, SOX 17), and renal interstitial cells (MEIS-1); cell nuclei were labeled with DAPI (blue color). Scale 50 μM. (Image courtesy of Takasato et al. 2014 [49])

organs at the early stages of embryonic development (during organogenesis) provides valuable information for precise diagnosis.

In recent years, organoid technology has been applied to in vitro generation of multicellular tissue constructs by recapitulating self-organizing properties of natural organogenesis. Great progress has been made in organoid technology; however, still there are several limitations in the generation of organoids (e.g., in size, thickness, structural complexity, and signal transduction) for clinical application.

In addition, natural organogenesis is a highly dynamic process that transduces thousands of signals among different cells per second. Although organoid technology possesses several advantages over conventional methods of tissue engineering in 3D cultures, it still lacks the complete maturation and functionality of normal tissue. Lack of total maturity of organoids reveals that microenvironmental signals are not dynamic enough or are not presented in the appropriate time and space. By a combination of methods of gene targeting, inducible gene expression methods, as well as stem cell technology, material science, and microfluidics technology, it is possible to produce highly effective organoids in the future.

Appropriate modeling of dynamic signal transduction is critical for maturation of organoid tissue constructs. To this end, application of conductive matrices that could freely transduce any chemical, electrical, and physical signals is critical. In addition, application of microfluidic systems can facilitate the dynamic delivery of signaling molecules. Application of stimuli-responsive materials can also accelerate signal transduction between cells and their microenvironment.

References

1. Adler R, Canto-Soler MV (2007) Molecular mechanism of optic vesicle development: complexities, ambiguities, and controversies. Dev Biol 305:1–13
2. Agathocleous M, Harris WA (2009) From progenitors to differentiated cells in the vertebrate retina. Annu Rev Cell Dev Biol 25:45–69
3. Barker N et al (2007) Identification of stem cells in small intestine and colon by marker gene Lgr5. Nature 449:1003–1007
4. Brabletz T (2012) EMT and MET in Metastasis: Where Are the Cancer Stem Cells?. Cancer Cell 22 (6):699–701
5. Bressan RB et al (2017) Efficient CRISPR/Cas9-assisted gene targeting enables rapid and precise genetic manipulation of mammalian neural stem cells. Development, 144(4), 635–648.
6. Dean DM, Napolitano AP, Youssef J, Morgan JR (2007) Rods, tori, and honeycombs: the directed self-assembly of microtissues with prescribed microscale geometries. FASEB J 21(14):4005–4012
7. Detrick RJ et al (1990) The effects of N-cadherin misexpression on morphogenesis in Xenopus embryos. Neuron 4:493–506
8. Eiraku M et al (2011) Self-organizing optic-cup morphogenesis in three-dimensional culture. Nature 472:51–56
9. Eiraku M et al (2012) Relaxation- expansion model for self-driver retinal morphogenesis: a hypothesis from the perspective of biosystems dynamics at the multi-cellular level. BioEssays 34:17–25
10. Esteve P, Bovolenta P (2006) Secreted inducers in vertebrate eye development: more functions for old morphogens. Curr Opin Neurobiol 16:13–19
11. Foty RA et al (1996) Surface tensions of embryonic tissues predict their mutual envelopment behavior. Development 122:1611–1620
12. Fuhrmann S (2006) Wnt signaling in eye organogenesis. Organogenesis 4:60–67
13. Fujimori T et al (1990) Ectopic expression of N-cadherin perturbs histogenesis in Xenopus embryos. Development 110:97–104
14. Gauvin R, Ahsan T, Larouche D, Levesque P, Dube J, Auger FA, Nerem RM, Germain L (2010) A novel single-step self-assembly approach for the fabrication of tissue-engineered vascular constructs. Tissue Eng Part A 16(5):1737–1747

15. Gilbert S (2000) Developmental biology, 6th edn. Sinauer Associates, Sunderland, MA
16. Gjorevski N et al (2014) Bioengineering approaches to guide stem cell-based organogenesis, the company of biologists. Development 141:1794–1804
17. Grayson WL, Martens TP, Eng GM, Radisic M, Vunjak-Novakovic G (2009) Biomimetic approach to tissue engineering. Semin Cell Dev Biol 20(6):665–673
18. Hadjimichael C (2015) Common Stemness regulators of embryonic and cancer stem cells. World J Stem Cells 7(9):1150–1189
19. Hindley C et al (2016) Organoids from adult liver and pancreas: stem cell biology and biomedical utility. Dev Biol 420:251–261
20. Humphreys BD (2014) Kidney structures differentiated from stem cells. Nat Cell Biol 16:19–21
21. Ishiwata T (2016) Cancer stem cells and epithelial-mesenchymal transition: Novel therapeutic targets for cancer. Pathology International 66 (11):601–608
22. Jabbari E (2011) Bioconjugation of hydrogels for tissue engineering. Curr Opin Biotechnol 22(5):655–660
23. Jakab K, Norotte C, Marga F, Murphy K, Vunjak-Novakovic G, Forgacs G (2010) Tissue engineering by self-assembly and bio-printing of living cells. Biofabrication 2(2):022001
24. Kachouie NN, Du Y, Bae H, Khabiry M, Ahari AF, Zamanian B, Fukuda J, Khademhosseini A (2010) Directed assembly of cell-laden hydrogels for engineering functional tissues. Organogenesis 6(4):234–244
25. Kobel S, Lutolf MP (2011) Biomaterials meet microfluidics: building the next generation of artificial niches. Curr Opin Biotechnol 22(5):690–697
26. Kretzschmar K, Clever H (2016) Organoids: modeling development and the stem cell niche in a dish. Dev Cell 38:590–600
27. Lancaster M, Knoblich JA (2014) Organogenesis in a dish: modeling development and disease using organoid technology. Science 345(6194):1–9
28. Little MH, McMahon MP (2012) Mammalian kidney development: principles, progress, and projections. Cold Spring Harb Perspect Biol 4:a008300. https://doi.org/10.1101/cshperspect.a008300 pmid: 22550230
29. Livoti CM, Morgan JR (2010) Self-assembly and tissue fusion of toroid-shaped minimal building units. Tissue Eng Part A 16(6):2051–2061
30. Lund AW (2009) The natural and engineered 3D microenvironment as a regulatory cue during stem cell fate determination. Tissue Eng Part B 15(3):371–380
31. Lutolf M, Blau H (2009) Artificial stem cell niches. Adv Mater 21:3255–3268
32. Mironov V, Visconti RP, Kasyanov V, Forgacs G, Drake CJ, Markwald RR (2009) Organ printing: tissue spheroids as building blocks. Biomaterials 30(12):2164–2174
33. Miranov V, Kasyanov V, Markward RR (2011) Organ printing: from bioprinter to organ biofabrication line. Curr Opin Biotechnol 22:667–673
34. Miyajiamn A et al (2014) Stem/progenitor cells in liver development, homeostasis, regeneration and reprogramming. Cell Stem Cell 14(5):561–574
35. Moscona A (1961) Rotation-mediated histogenetic aggregation of dissociated cells. A quantifiable approach to cell interactions in vitro. Exp Cell Res 22:455–475
36. Oh J et al (2014) Stem cell aging; mechanisms, regulators and therapeutic opportunities. Nat Med 20(8):870–880
37. Park D et al (2015) Stem cell microenvironment on a chip, current technologies for tissue engineering and stem cell biology. Stem Cell Transl Med 4:1352–1368
38. Rodriguez- Seguel E et al (2013) Mutually exclusive signaling signatures define the hepatic and pancreatic progenitor cell lineage divergence. Genes Dev 27(17):1932–1946
39. Rossi JM et al (2001) Distinct mesodermal signals, including BMPs form the septum transversum mesenchyme are required in combination for hepatogenesis from the endoderm. Genes Dev 15(15):1998–2009
40. Rothermel A et al (1997) Pigmented epithelium induces complete retinal reconstitution from dispersed embryonic chick retinae in reaggregation culture. Proc Biol Sci 264:1293–1302

41. Sato T, Clevers H (2013) Growing self-organizing mini-gut from a single intestinal stem cell: mechanism and applications. Science 340:1190–1194
42. Sato T et al (2009) Single Lgr5 stem cells build crypt-villus structures in vitro without a mesenchyma niche. Nature 459:262–265
43. Schultz MB, Sinclair DA (2016) When stem cells grow old: phenotypes and mechanisms of stem cell aging. Development 143:3–14
44. Steinberg MS (1964) In: Locke M (ed) Cellular membranes in development. Academic Press, New York, pp 321–366
45. Sweeney P et al (2017) Protein misfolding in neurodegenerative diseases: implications and strategies. Transl Neurodegeneration 6:1–13
46. Takebe T et al (2013) Vascularized and functional human liver from an iPSC-derived organ bud transplant. Nature 499:481–484
47. Taguchi A et al (2014) Redefining the in vivo origin of metanephric nephron progenitors enables generation of complex kidney structures from pluripotent stem cells. Cell Stem Cell 14:53–67
48. Takasato M, Little M (2016) A strategy for generating kidney organoids; recapitulating the development of human pluripotent stem cells. Dev Biol 420:210–220
49. Takasato M et al (2014) Directing human embryonic stem cell differentiation towards a renal lineage generates a self-organizing kidney. Nat. Cell Biol. 16:118–126
50. Valastyan J, Lindquist S (2014) Mechanisms of protein folding diseases at a glance. Dis Model Mech 7(1):9–14
51. Weiss P, Taylor AC (1960) Reconstitution of complete organs from single-cell suspensions of chick embryos in advanced stages of differentiation. Proc Natl Acad Sci U S A 46:1177–1185
52. Wolpert L (2007) Principles of development, 3rd edn. Oxford University Press, Oxford
53. Xia Y et al (2013) Directed differentiation of human pluripotent cells to ureteric bud kidney progenitor-like cells. Nat. Cell Biol 15:1507–1515
54. Yabut O, Bernstein H (2011) The promise of human embryonic stem cells in aging-associated diseases. Aging 3(5):494–508
55. Yin X et al (2016) Engineering stem cell organoids. Cell Stem Cell 18:25–37
56. Žigman M et al (2005) Mammalian inscuteable regulates spindle orientation and cell fate in the developing retina. Neuron 48:539–545
57. Zhang Z et al (2017) CRISPR/Cas9 Genome-Editing System in Human Stem Cells: Current Status and Future Prospects. Mol Ther Nucleic Acids, 9, 230–241.

Chapter 7
Conclusion

In recent years, information theory has become the focus of research on several issues ranging from quantifying and understanding autonomous systems to questions of spontaneity in biology, robotic behavior, and artificial intelligence in technology, as well as the generation of smart, programmable materials. However, autonomy is still a major challenge to be explored. In this book, we attempted to describe the regulatory mechanisms behind autonomous properties of the biological system by connecting the dots across different areas including endogenous coding, cognition, and adaptation with environmental conditions, structural flexibility, as well as self-organization.

To this end, firstly we referred to embryonic organogenesis as a promising model system for an understanding of regulatory mechanisms behind the self-organization and autonomy in biological systems.

Biological systems are defined as advanced programmed systems due to the capability to store information in their molecules. Nucleic acids are well-known to store and process information (code) in biological systems. DNA is a natural molecule designed by nature over billions of years of natural evolution as an information storage media. In addition to the capability of nucleic acids to store nature's information, recent research in the field of DNA computing extended the capability of DNA molecules to computer science by solving highly complicated mathematical problems including numerical optimization problems (NP) (as discussed in Chap. 2). Therefore, here in this book, we attempted to elaborate on this unique aspect of nature's design to apply molecules as a matrix for information (code) storage.

Our current understanding of nature's laws in physics and chemistry is limited to the relativity of energy and mass. As explained in this book, nature also has thousands of different types of molecules as a matrix for information storage (other than energy storage). Nature applies highly efficient and sustainable methods by designing of multifunctional materials that can store both information and energy in their molecules. Therefore, we considered the possibility of information storage in molecules as an additional dimension of nature's law, defining the new concept of relativity of code, energy, and mass. The relativity of code, energy, and mass can explain

T. Karimi, *Molecular Mechanisms of Autonomy in Biological Systems*,
https://doi.org/10.1007/978-3-319-91824-2_7

autonomous properties of biological systems from different perspectives, including endogenous coding, cognition, decision-making, dynamic adaptation with environmental conditions, as well as self-regeneration and self-fueling capacities.

We considered the human body as an autonomous system that has been precisely programmed by a molecular coding system. The coding system of the human body is not limited to the genetic code (or DNA coding system). But, the coding system of the human body is made from multiple layers of coding molecules including DNA, mRNA, tRNAs, proteins, hormones, neurotransmitters, etc. Therefore, all physiological functions of the human body are regulated through a highly harmonic multilayer coding system with all of its components precisely connected to each other. Multifunctional properties of biomolecules lead to the harmonic connection between different compartments of this molecular coding system. In Chaps. 5 and 6, we explained how relativity of code, energy, and mass orchestrates all operations in biological systems in a completely autonomous manner. Keeping in mind that the whole human body is a highly programmed system, any diseases can be considered as miscoding or a disorder that originated from different layers of this coding system in the human body. In Chap. 6, we attempted to shed light on the organogenesis as a precisely programmed process where all tissues and organs of the human body initially formed at early stages of embryonic development. Deep understanding of the molecular mechanisms of organogenesis through the molecular signaling networks can open new avenues toward the treatment of diseases incurable with the conventional biomedical approaches including cancer, aging, and degenerative diseases. In near future, by application of basic principles of the relativity of code, energy, and mass in designing smart, programmable materials, we can provide a solution to correct the misfunctions correlated to different diseases.

Index

T. Karimi, *Molecular Mechanisms of Autonomy in Biological Systems*,
https://doi.org/10.1007/978-3-319-91824-2

MIX
Papier aus verantwortungsvollen Quellen
Paper from responsible sources
FSC® C105338

If you have any concerns about our products,
you can contact us on
ProductSafety@springernature.com

In case Publisher is established outside the EU,
the EU authorized representative is:
Springer Nature Customer Service Center GmbH
Europaplatz 3, 69115 Heidelberg, Germany

Printed by Libri Plureos GmbH
in Hamburg, Germany